ORGANIZE OR BURN

Organize or Burn

How New York Socialists Fight for Climate Survival

Fabian Holt

NEW YORK UNIVERSITY PRESS
New York

NEW YORK UNIVERSITY PRESS
New York
www.nyupress.org

© 2025 by New York University
All rights reserved

Please contact the Library of Congress for Cataloging-in-Publication data.

ISBN: 9781479837809 (hardback)
ISBN: 9781479837816 (paperback)
ISBN: 9781479837830 (library ebook)
ISBN: 9781479837823 (consumer ebook)

This book is printed on acid-free paper, and its binding materials are chosen for strength and durability. We strive to use environmentally responsible suppliers and materials to the greatest extent possible in publishing our books.

The manufacturer's authorized representative in the EU for product safety is Mare Nostrum Group B.V., Mauritskade 21D, 1091 GC Amsterdam, The Netherlands. Email: gpsr@mare-nostrum.co.uk.

Manufactured in the United States of America

10 9 8 7 6 5 4 3 2 1

Also available as an ebook

There was a certain critical period. I see that now. During that period, your grandmother and I were doing, every night, a jigsaw puzzle each, at that dining-room table I know you know well, we were planning to have the kitchen redone, were in the midst of having the walls out in the yard rebuilt at great expense, I was experiencing the first intimations of the dental issues I know you have heard so much (too much?) about. Every night, as we sat across from each other, doing those puzzles, from the TV in the next room blared this litany of things that had never before happened. . . .

So, although your grandmother and I, during this critical period, often said, you know, "Someone should arrange a march" or "Those f——ing Republican senators," we soon grew weary of hearing ourselves saying those things. . . . We were not prepared to drop everything in defense of a system that was, to us, like oxygen: used constantly, never noted. We were spoiled, I think I am trying to say. . . .

What would you have had me do? What would you have done? I know what you will say: you would have fought. But how?

—George Saunders, "Love Letter" (2020)

CONTENTS

Introduction: A Movement Party in the Climate Crisis	1
1. The Climate Movement Needs Politics!	31
2. The Sociology of Movement Parties	47
3. The Political Environment After the Great Recession	63
4. Anticommunism Kills	80
5. A Departure from the Democratic Party in Brooklyn	91
6. The Alexis Campaigners in Flatbush	112
7. Organization and Leadership in NYC-DSA	143
8. The Party Dimension	178
9. The Road to Green New Deal Legislation	192
Conclusion 1: Status of the Democratic Road to Climate Survival	229
Conclusion 2: Scholarship for Survival	237
Acknowledgments	245
Notes	247
References	267
Index	287
About the Author	299

Introduction

A Movement Party in the Climate Crisis

Some say we should not engage in activism. Instead we should leave everything to our politicians and just vote for a change instead. But what do we do when there is no political will? What do we do when the politics needed are nowhere in sight?
—Greta Thunberg (2019)

As a college professor, I am witnessing, every year, how more young people are becoming concerned about climate change. However, many of my students still avoid the topic because they find it uncomfortable. Yes, you read that right. A student said to me the other day, "I wish we had been assigned a different topic because climate change is kind of depressing." I teach at a university in Denmark, which has thus far had no catastrophic climate hazards. The student's sentiment is common. It is one of the most popular forms of climate apathy, premised on the hope that the problem will be fixed by someone or something else. Perhaps the next generation? A technological breakthrough? Meanwhile, global greenhouse gas emissions continue to rise. What are the consequences? Decades of climate inaction are already the cause of widespread suffering and devastation. By conservative estimates, eight hundred thousand people are dying every year from causes related to global warming, and millions are forcibly displaced, the majority of them in low-latitude regions such as Africa, South Asia, and the Caribbean. Images of floods and wildfires circulate widely, but many people die silently and invisibly in heat waves. Those of us who live on experience more severe climate hazards and weakening global ecosystems, resulting in fractured communities, health crises, "heatflation," and material damage that can no longer be insured.[1] That is why more people are rethinking their priorities and wrestling with the basic questions of climate politics. Who is responsible for our climate future? Why are politicians not doing more to save us? Why is the fossil fuel industry still expanding when renewable alternatives are available? Why are poor people and developing countries suffering disproportionally? What are the underlying problems of our current political system, why is it so difficult to hold political institutions accountable, and how can those problems be fixed?

This book is about a movement that responds to these big questions by emphasizing basic principles of democracy and social justice. The movement contests in elections to gain political influence but without aspiring to become an ordinary political party. It seeks to correct some of the problems in political institutions that are widely acknowledged but seem so difficult to fix, and the movement is thus paradoxically perceived as a boundary phenomenon. More specifically, this book is about the particular New York variant of the international democratic socialist movement. In the wake of the Great Recession of 2008, movements in several countries emerged that revived the concept of "democratic socialism." Developed as an alternative to communist Marxism in the early 1900s, democratic socialism is about using democratic action, in social movements and government, to create a more equal society, with a more equal distribution of income flows. Democratic socialism fundamentally defined market-based welfare societies in the twentieth century and is now being adapted to the Anthropocene.

This book tells the untold story of how young people in the epicenter of the democratic socialist movement in the United States developed a distinct political agenda and organizational approach to the climate crisis. The book argues that both aspects—the political project and the approach to political organizing—are important to repairing democracy and overcoming the climate crisis.

The subject of political organizing has become urgent because the ecological crisis has evolved into a political crisis. Governments have expressed concern and defined long-term goals, but they are not meeting their goals. Ten years on, none of the forty leading economies that agreed to the Paris Agreement in 2015 are on track to do what they promised. Climate has become a matter of information management and expert administration, framed by the discourse of corporate environmentalism that a network of public relations agencies engineered in the 1990s. Governments have played along by setting up frameworks for monitoring climate change and supported the expansion of fossil fuel production, thus articulating the function of the neoliberal state as an enabler of market growth. As a matter of fact, governments have not taken the most obvious actions to protect society, such as phasing out fossil fuels, investing in public transportation, and informing the public about risks and solutions. Governments have banned advertising for smoking and alcohol and run information campaigns. Why are they not doing the same for climate? The misalignment of governments with nature, science, and the public is developing into what communication scholars call a "sticky" crisis: a long-lasting crisis that seemingly elides resolution and therefore becomes increasingly intense and harmful. More citizens are experiencing a fundamental sense of injustice and neglect. Why are governments siding with fossil fuel companies instead of environmental organizations? Why are climate activists being put to jail, while oil companies are making billions?[2]

How can this situation be turned around? Some commentators suggest that citizens should start "saving themselves." I agree that citizens should not wait for institutions to act, but the societal transformation required to avoid climate collapse cannot be achieved through individual behavioral change or protest movements alone. The tradition of democratic socialism combines movements with electoral politics to build popular support and not simply take down elite rulers. Authoritative political institutions are necessary for social justice. By contrast, communist thinkers such as Karl Marx and Vladimir Lenin mandated the working class to overthrow the capitalist system by violent means if necessary and eventually dissolve the state. Contemporary democratic socialism is partly a response to authoritarian right-wing movements.[3]

The state can play a unique and necessary role in the climate crisis because of its monopoly of legitimate power to create and enforce laws. The state authorizes the social order and institutes moral norms. It can use that power to protect current and future generations. It can create administrative justice in the form of bureaucracy. Only the state can effectively regulate industries, guarantee justice for vulnerable populations, and transform energy and transportation infrastructures that are not profitable in the short term.[4]

Thus, from a democratic socialist perspective, ordinary people should not give in to climate apathy but organize in movements and parties to make the state work for their collective climate future. Without a stronger democracy and more socially just policies, we are delegating power to autocratic leaders, fossil fuel companies, and disaster capitalism.

The democratic road to climate survival is complicated by a systemic crisis of democracy, however. The crisis goes beyond familiar issues such as spin, dark money, and career politicians. Democracies have regressed into authoritarian and fascist neoliberalism, epitomized by the likes of Donald Trump in the United States, Jordan Bardella in France, and Giorgia Meloni in Italy. These leaders promote nativism and racism instead of democratic principles of egalitarianism and pluralism. They show disregard for human life, the truth, and government and transform democratic states into mafia states by making personal loyalty more important than anything else.[5]

The crisis of democracy has accelerated in neoliberalism, which began as the political project of reestablishing elite power during the recession in the 1970s. Social democracy—the most popular form of democratic socialism in history—was blamed for declining growth and rising inflation. To reestablish the conditions for capital accumulation, governments abandoned Keynesian ideas of state intervention in the market and lowered top tax rates, privatized public assets, and deregulated international trade and market exploitation, while cutting welfare and weakening labor unions. England and the United States took the lead and coerced developing countries to comply via institutions such as the International Monetary Fund and the World Bank. This epochal shift in politics was related to broader changes in society. Social democratic values lost traction with the declining number of manufacturing workers, and neoliberalism resonated with growing individualism. The societal consequences are tangible, with four decades of rising global economic inequality and the erosion of popular democracy, which makes political institutions more vulnerable to authoritarianism. Rising inequality, global warming, and geopolitical issues culminated in the 2020s in an atmosphere of dystopia and polarization. I began writing this book in New York City in 2022, when half of the population was exhausted by the pandemic and recovering from the trauma of Trump's first term. Trump had openly disrespected government and science, validated racism, lied, denied the existence of a deadly disease, and even encouraged violent insurrection. Trump has now returned for a second term, vengeful and reckless.

The combination of neoliberalism and Trumpism amplifies a decades-old vicious cycle of political participation and democratic decline. This is a cycle in which bad politics leads to more indignation, to less participation, and ultimately to worse politics. Climate change is caught in this vicious cycle and takes it to new extremes.[6]

The Insurgency of Socialist Movement Parties

This book examines how thousands of young New Yorkers have reinvented democratic socialism to take on the major challenges of the twenty-first-century city. Motivated by broader changes in the political environment, these young New Yorkers joined the Democratic Socialists of America and updated its political project and methods. With the rise of authoritarianism in the 2016 election as a catalyst, the New York City chapter of the Democratic Socialists of America (NYC-DSA) developed into an urban demos and laboratory of democracy. It also became the leading force in New York State's transition toward renewable energy and provides an instructive example of how climate activism can be channeled into electoral politics. This book gives voice to people who were most active in NYC-DSA's transformation in the late 2010s and examines their successes and failures with perspectives for the broader conversation around climate and democracy.

But first, let us recognize the international revival of democratic socialism during the Great Recession of 2008. As governments cut welfare and bailed out banks, and banks paid out billions in bonuses, millions of people fell into poverty and lost their homes to foreclosure. The Great Recession triggered a new skepticism about capitalism, especially among young people who had not lived through the Cold War. Naomi Klein was a generation older and remembered in 2022 that the Cold War stigma of socialism "had rooted itself so deeply in the collective imagination that, even when many of us did start defining ourselves as anti-capitalist, we remained fearful, for far too long, of articulating a coherent vision for a postcapitalist world."[7]

In southern Europe, the socialist revival first took the form of protest movements, and some of these movements evolved into parties to gain influence and institutional representation. These parties initially maintained skepticism toward established parties and sustained relationships with their movement base, illustrating different approaches to the hybrid phenomenon of movement parties. Ideologically, the democratic socialist parties define themselves as a response to liberal and social democratic parties moving right. Compared to green parties, they have a more universalist political program and demand a more radical transformation of capitalism. In the US, democratic socialists have accomplished something that the Green Party has not, namely, mobilizing a popular climate Left in the Green New Deal movement. They have politicized the democratic and economic dimensions of climate change. Today, the slogan "People over profit" can be heard at climate rallies everywhere, and that is just the tip of the iceberg, as we shall see. However, the climate Left is facing new headwinds in the 2020s from neoliberal fascist movements and the general sidelining of climate interests in business and politics. Enthusiasm for climate as the next frontier of economic growth has faded, corporations are backing away from their climate targets, and governments are channeling massive resources into militarization.[8]

The democratic socialist parties demonstrate the relevance of movement party organizations to climate survival. These organizations combine the party elements of having a political program, running candidates for elections, and participating in the institutional sphere, with the movement elements of grassroots democracy and protest in extrainstitutional contexts outside the polity and beyond the boundaries of mass parties. A key point of this book is that a democratic socialist movement party is not just an organizational design but also a set of values and practices. It is an attempt at reclaiming democracy under its newly challenged conditions, crucially involving collaboration across movement and institutional contexts on the basis of shared values.

The movement party phenomenon originates in the transformation of national political fields across Europe in the 1960s and 1970s. During this time, a decades-old system of the same mass parties was destabilized. Old parties began to lose, their traditional constituents declined, and new protest parties quickly became big in so-called earthquake elections in the 1970s. Movements grew when parties started to lose. They expanded the political realm into new subjects and topics. In the process, movements broke the monopoly of parties as vehicles for political socialization, organization, and mediation between citizens and political institutions. As parties have increasingly become "hollowed out" in neoliberalism, relying more on elite networks and donations rather than grassroots participation, organic grassroots movements have gained renewed relevance in the twenty-first century.[9]

The movement party phenomenon emerged in the 1980s when ecological movements developed into green parties to gain influence, representation, and institutional stability. The green parties illustrate the core potential of movement parties, namely, the expansion from a single-issue agenda and short-term organization for protest into a broader political program and a more permanent organization. This works if supporters find that the salient issues are codependent and can therefore be meaningfully connected in a more general political program. In a climate context, for instance, the perspective can evolve from pipeline protests to the development of climate policy with provisions for the energy sector and worker retraining. Such development requires considerable resources and involves a lot of trial and error. Movement parties are frequently a transitory stage of political organization, because once the party is created, it adapts to the new institutional environment of other parties.[10]

In the twenty-first century, movements remain symptomatic of the declining capacity of political parties to sustain linkages with citizens. They are now also symptoms of a broader crisis of democracy and the migration of politics into social media. Political parties increasingly rely on movements fueled by indignation about political institutions, but there are crucial differences between movement dynamics on the Right and the Left.

On the far Right, movements rally in support of autocratic leaders. Ordinary people can be fans and supporters, but there is little grassroots democracy. Leaders appeal to a feeling of movement belonging and encourage participation in rallies, protests, and even insurrection. In the United States, Tea Party movement leaders have entered the Republican Party to "movementize" the party within its institutional context. In

France and Germany, right-wing parties have similarly been movementized and voiced antigovernment sentiments. Populist right-wing parties generally receive large amounts of dark money and are influenced by economic elites seeking tax reductions. In the US, much of the Tea Party's visibility came from above, from large and wealthy advocacy groups that found an outlet to claim authorship of a movement and promote tax reductions and the privatization of health care.[11]

Leftist movements interpret the problem with political institutions in terms of inequality and democratic failure. They seek to revitalize democracy by recruiting more diverse candidates and rejecting corporate donations. They are skeptical of oligarchy and see movement parties as a path for overcoming the underlying structural issue in the crisis of democracy, namely, the widened social cleavage between the economic elites and ordinary people in neoliberal capitalism. This has implications for how democratic socialists understand the climate crisis. They argue that the interlocking crises of inequality and democracy are blocking the ecological transformation of society. They do not place faith in ad hoc solutions, such as referendums and citizen assemblies. Democratic socialists object that such initiatives do not build a broad cultural and institutional base of democracy or fix the underlying structural problem of inequality. After all, why would politicians relying on funding from economic elites delegate power to citizen assemblies? Citizens assemblies and other participatory processes are often a form of tokenism in the neoliberal order.[12]

Why New York City?

The crisis of democracy is particularly severe in the United States. Democracy has long been constrained by a complex voter registration system, party duopoly, and campaign finance laws skewed toward the economic elites. The current crisis of democracy is more cultural and evolved from the Christian nationalist movement. In 2023, *The Economist*'s international democracy index gave the US a score of 7.85, categorizing it as a "flawed democracy." The rise of authoritarianism might seem like an opportunity for democratic socialism, but the two-party system remains an obstacle and explains why the socialist movement has not developed into a party the United States.

Socialists have been forced to develop creative solutions in the United States, and this why New York City is central: the concentration of cultural resources.¹³

New York City is where democratic socialism gained critical mass, and the city's history illustrates broader trends. Long into the twentieth century, New York was a working-class city and a social democratic polity. It was committed to racial equality and welfare policies of access to education and culture. It institutionalized rent control and public housing when the rest of the country eliminated those things. Unions were strong and created a bulwark against anticommunist propaganda. However, the class demographics of Manhattan changed in the 1950s and 1960s, with the growing dominance of white-collar workers, many of whom were employed in new headquarters of global corporations. Two million people left the city between 1940 and 1960. During the fiscal crisis in the 1970s, the economic elites grabbed power and undid many of the historical achievements of the working class.¹⁴

Thirty years later, a radical anticapitalist opposition began with the Occupy Wall Street encampments in 2010 in response to the Great Recession. A democratic socialist insurgency evolved at the national level and inspired the creation of a movement party in the New York City chapter of the Democratic Socialists of America, the largest and most influential organization for democratic socialism in the country.

The "new" NYC-DSA emerged in Brooklyn as an alternative to the traditional and semicorrupt "machine" politics of the Democratic Party organization there. It also articulates broader changes in the economics, geography, and communication systems of political life. NYC-DSA is translocal, queer, multiracial, and digitally networked. It has broad relevance as a laboratory of contemporary political life and alternatives to the dominant protocols of liberal democracy.¹⁵

Political scientists have long considered New York State as an influence on the United States' constitutional system and a laboratory of political administration since its first constitution, the Articles of Confederation, was adopted in 1777. But the state's political life also has a history of grassroots politics, and NYC-DSA is a new chapter in this history.¹⁶

The "New" NYC-DSA

How did the socialist revival evolve more specifically in the United States and New York City? The Great Recession spread from the United States to fifty-nine countries at its peak in 2009. A key factor was the deregulation of financial institutions, making high-risk lending possible. Insiders described Wall Street as a place of greed and misplaced optimism. The historian Eric Hobsbawm commented a few years later that the world of global capitalism uncannily resembled the world anticipated by Marx in the *Communist Manifesto*—a world of exploitative class struggle, ruthless economic self-interest, social fragmentation, and compromised working-class dignity.[17]

New York City was also the birthplace of an influential movement response to the Recession. The Occupy Wall Street encampments of 2010 introduced anticapitalist narratives that triggered broad changes in political life. Many experienced leftists observed how the Occupy narratives resonated with young people. One of them was Senator Bernie Sanders of Vermont, a lifelong socialist and independent politician. He made the bold move in 2015 to run an insurgent presidential primary campaign on an explicitly socialist agenda and won thirteen million votes the following year.[18]

The Sanders campaign in 2016 created temporary structures for a democratic socialist movement, but they were temporary. Sanders created the organization Our Revolution to sustain momentum after his first thunderous run and to reform the Democratic Party organization. Our Revolution had little success and did not build organizational capacity for the movement.[19]

The torch from Sanders was picked up by a new generation who joined the Democratic Socialists of America (DSA). DSA is a decentralized organization concentrated in large cities, and the most extraordinary transformation happened in New York City, where it was able to achieve critical mass. New members flocked to large assemblies in Brooklyn, some of which had six hundred people in attendance. They transformed the city's DSA chapter (NYC-DSA) from a political club into an organization of collective action and an informal party. A small network of people had paved the way. They had experience in community and labor organizing and were inspired by new socialist media outlets such as *Jacobin* magazine, which emerged from DSA's network in New York City. *Jacobin* staff organized reading groups across the country and eventually helped them transform into DSA chapters. The reading group in Brooklyn was popular and included Julia Salazar, who won NYC-DSA's first seat in state government in 2018.

Some of the new young organizers in NYC-DSA in 2016 came from a tiny club called the Brooklyn Socialist Club, which was founded in 2014 and had just about twenty active members. Sam Lewis was one of the organizers in this club and remembers that they took inspiration from the 2013 victory of a candidate for city council in Seattle named Kshama Sawant of the revolutionary Socialist Alternative Party. After Sanders announced his run in 2015, these emerging young organizers started to feel that socialism had more potential and backed Debbie Medina, a housing activist, who ran for the state senate in 2016, saying, "I'm a democratic socialist."

Donald Trump's election in November 2016 motivated thousands to join NYC-DSA. Several of the first big meetings in 2016 took place at the Mayday Space in Bushwick, a community center and grassroots organizing hub led by several former Occupy activists. That is where the idea to form an electoral working group began, and the proposal was formally passed at a meeting at the Brooklyn Free School in Clinton Hill, also in northwest Brooklyn. The organizers created spreadsheets of fifty locations and their capacities and costs and contact persons, because the organization did not have an office. NYC-DSA opened its first office on the Lower East Side in Manhattan in 2020 and planned a second office in Brooklyn's Bedford-Stuyvesant in 2025. Since 2017, it has had a regular presence in its electoral campaign offices, however. Sam Lewis remembers the time in late 2016:

> LEWIS: Everyone joined DSA because they were, like, "We needed a political response to Trump that is not Clinton, you know, neoliberal." But it was not clear what we were going to do. We had some big meetings of six hundred people, and we were asking, "What are we doing?" It's lucky that the few of us who had been involved in DSA and the club with the Debbie Medina campaign came in and said, "We can recruit, endorse, and win local socialist campaigns and build this political movement that Bernie started. That was the start of the

Electoral Working Group. By the time that rolled around, I felt like, "We have a plan," a very open-ended one. My ideology and political strategy are now much more fixed, but at that time, we were like, "We should do whatever works. We should see what happens!"

AUTHOR: And then other DSA chapters around the country started doing electoral politics?

LEWIS: That's right. A lot of chapters did. In 2018, when AOC [Alexandria Ocasio-Cortez] and Julia [Salazar] won, that was a big lightbulb across the country that, like, "There's a future for Bernie politics beyond Bernie Sanders in 2016." The AOC upset was earth-shattering, and it was lucky that Julia won right after that.[20]

The organization developed quickly. NYC-DSA has won elections in more districts and now has nine electeds in the New York State government, which is headquartered in Albany. The electeds represent about 1.8 million constituents, and one of them—Zohran Mamdani—became a household name in New York City when he ran for mayor in the 2025 election. The organization has expanded from one citywide branch in 2016 to eight neighborhood branches, spanning a distance of twenty-five miles from Staten Island and Bay Ridge in the south to the Bronx and Queens in the north. New thematic working groups have emerged, including the Ecosocialist Working Group, which has become a transformative force within the organization and spearheaded the first Green New Deal legislation in the country, the New York State Build Public Renewables Act of 2023. What is more, NYC-DSA has diversified culturally. Leaders increasingly draw inspiration from racial and climate justice movements. By 2020, it was no longer dominated by white male followers of Sanders. The majority of the four thousand who joined in the months after October 7, 2023, were people of color.[21]

DSA was originally founded in New York City in 1982 out of the ashes of the young "New Left" of the 1960s, which burned out during the 1970s from factional struggles and radicalization. DSA defined itself as an alternative to the far Left and rejected Stalinism. Its leaders, including co-chairs Michael Harrington and Barbara Ehrenreich, were prolific and influential writers. Many members participated in political activism and campaigns, but DSA did not have a focus on political organizing. By the 2000s, it had become a somewhat sleepy discussion club. DSA's transformation began in New York City as part of the wider national anti-Trump movement, but its focus extended beyond protest. The young people who joined DSA were inspired by Sanders to sustain a socialist movement and transform the Democratic Party.

In 2013, DSA had six thousand members with an average age of sixty-seven. The number climbed from thirty-two thousand in 2017 to ninety-five thousand in 2021, and the average age dropped to thirty-three. The COVID-19 pandemic generally exhausted civil society organizations and contributed to a decline at DSA, but the organization reversed the trend in 2023. In 2023, DSA's national membership stood at seventy-eight thousand. In New York City, the DSA had six thousand members and a revenue of $480,000, mostly from membership fees and individual donations for campaigns such as Palestinian Solidarity, Tax the Rich, and Union Power; 7,679 New Yorkers participated in the "No Money for Massacres" campaign. NYC-DSA thus still has the critical mass and organizational capacity to develop and run movement pressure campaigns and electoral campaigns. This is "the new NYC-DSA": a young movement-based organization with a party dimension shaped by the post-Recession socialist and climate movements.[22]

While the transformation of NYC-DSA might seem like an isolated, local incident, it represents both typical and unique aspects of international movement responses to societal, political, and ecological crises that have been under way for decades. The millennials who built the new NYC-DSA were experiencing downward mobility. They were indignant about the failure of political elites to respond to accelerating economic inequality and climate change. They felt that the election of an autocratic president and the growing elite dominance in the Democratic Party signaled a deeper crisis of democracy. Like international socialist movement parties, NYC-DSA opposed elite influence in politics and favored grassroots democracy. DSA is the only organization of the anti-Trump movement that adopted a democratic socialist movement party approach, and this partly explains why DSA still exists, while many of the others are gone.

The story of NYC-DSA is also the story of the Brooklyn Democratic Party organization and New York City as a national center of socialism. For more than a hundred years, Brooklyn has been one of the country's biggest Democratic counties and an influential site of organized urban political life. Over the past decade, however, the Democratic Party organization has abandoned grassroots democracy, causing great frustration among young progressives in the party's support base. NYC-DSA offers a more democratic alternative and has challenged the situation in which the party can safely run passive campaigns. Its grassroots candidates have ousted longtime incumbents. The day after Julia Salazar's election in 2018, NYC-DSA's inaugural victory in the state government elections, *The New York Times* wrote that the national progressive wave had "hobbled New York's once-mighty Democratic machine." Centrist elected officials are now looking over their shoulders for a leftist primary challenge. NYC-DSA uses the Democratic Party's ballot to reach a broader electorate than would be possible if it were a third party. In this respect, it is an independent faction of the party.[23]

NYC-DSA is best understood as a movement-based informal party, however. It is a grassroots movement that also participates in elections. The purpose of the movement dimension is to create a grassroots sphere of politics with organizational capacity outside institutions. It is primarily mobilized through electoral campaigns and thus also has a party function. The movement dimension extends to the role of NYC-DSA electeds in state government. They not only are representatives of a party but also collaborate with the movement to win legislation despite the small number of socialists in government. The small movement and the small number of electeds thus collaborate to punch above their weight. NYC-DSA's movement party tactics have helped the organization move the needle in the Democratic Party and influence tax and climate legislation, for instance. These achievements should be understood as "movement achievements" won in coalition with organizations such as the Working Families Party, New York Communities for Change, and the Sunrise Movement.

I am writing this book because I believe, based on ethnographic experience, in the potential demonstrated by this organization, but I make no pretense. There are no magic solutions to the crises of democracy and climate change, but NYC-DSA demonstrates important potential, and this book seeks to build upon it.

Can we expect NYC-DSA to have built a mass movement and transform the Democratic Party in ten years? To offer complete solutions to the major crises of our time? Solutions that can be implemented elsewhere without much effort? Such expectations are wildly unrealistic. Broader political change from below requires long-term efforts across movements and parties to strengthen the connection between institutions and the everyday experience of ordinary people.

I understand why some of my colleagues are fundamentally skeptical about electoral politics. Two experienced labor movement historians, Jane McAlevey and Kim Moody, have argued that there is little potential for meaningful change through electoral politics. Moody argues in *Breaking the Impasse* (2022) that congressional politics is paralyzed by both major parties turning right and becoming dominated by elite interests over the past couple of decades. His analysis is insightful but also suffers from determinism. He is right that the Democratic caucus in Congress is dominated by wealthy donors and has tamed insurgent leftists such as AOC. But what if more people had voted for Sanders in 2016? What if the Democratic Party had selected its presidential candidate in 2024 democratically and that candidate had adapted less to the interests of wealthy donors? The Kamala Harris approach of posing with rich Hollywood celebrities and offering little of political substance surely represent the decay of an organization and not its future. What if more progressives and socialists win in the next election? There is no power structure that cannot be changed. Major changes usually happen in times of upheaval, and we are now in such a moment.

Reckoning with the Stigma of Socialism

The revival of democratic socialism results from a change in public consciousness and a new perspective on political history. Accelerating inequality has changed how many people think about capitalism. This aspect can be understood through the concept of "perspective transformation," which was introduced by the psychologist Jack Mezirow in the 1970s. The idea was developed for individual psychology, but it can be adapted to collective psychology. Mezirow's theory is that a person's exposure to new information or experiences can lead them to view a situation from a new and more problematic perspective and ultimately change their beliefs. Such a transformation is common during life crises. It is a process that reshapes how a person evaluates new experiences. Mezirow writes, "When a meaning perspective can no longer comfortably deal with anomalies in a new situation, a transformation can occur.... A new meaning perspective has dimensions of thought, feeling, and will.... Feelings and events are interpreted existentially, not intellectually as by an observer."[24] The material reality of income inequality partly explains the appeal of Sanders and the word "socialism" in 2016. In the words of a headline in *The New Yorker*, "Reality Has Endorsed Bernie Sanders." Inequality rose further during the pandemic, and even shareholders are now frustrated with executive pay! The socialism that focuses on everyday life and democracy has gained relevance, while utopian and revolutionary socialism remains obscure. The socialist revival is focused on practical organizing for basic social justice in the face of extreme inequalities.

The revival stands in the tradition of democratic socialism that developed in the labor movements of northwestern Europe in the late nineteenth century. Democratic socialism found its first expression in Eduard Bernstein's (1850–1932) "revisionist" thinking, which opposed core elements of Marxism. Bernstein did not believe that capitalism would collapse and argued that greater social justice should be achieved by strengthening democracy. He rejected revolutionary violence, arguing that the most important task was to build popular support democratically. The difference between democratic and revolutionary socialism became more distinct after the Russian Revolution in 1917. The revolution set out to create a stateless communist society of public ownership, but Russia and the other countries in the international communist movement of the mid-twentieth century all devolved into totalitarian regimes with party monopoly and practically abandoned communist ideology.

The main inspirational model for the socialist revival is the Nordic welfare state that provided citizens rights to material resources, such as universal health care, housing, education, and social insurance. Sanders brought these issues back into the popular political discourse in the context of extreme economic inequality, which was the foundational impetus for socialism in the nineteenth century.[25]

I was influenced by this transformation in public consciousness in the field. The experience changed my understanding of socialism. I grew up in the 1970s and 1980s with politically active parents committed to the Nordic welfare society model. My mother was active in

Denmark's leftist feminist movement, and my father was a member of the editorial staff at the country's foremost socialist magazine, *Politisk Revy* (Political review) in the late 1960s. I am named after the British Fabian Society, which pioneered democratic socialism, not the 1950s pop star Fabian Forte. However, neoliberalism had affected me in ways that I did not fully recognize until I participated in this movement in Brooklyn. I confess to having reservations when I first heard young New Yorkers talking about socialism and capitalism. I initially focused on gaps in their knowledge when they used these big words. This is a typical fallacy among academics and political experts. If we want to understand the agency and potential of ordinary people in politics, we need to understand how they think and feel about the world. Yet my journey, I would come to realize, was aligned with that of many NYC-DSA members in a particular way. I had been researching independent organizations in the cultural field for two decades and was now undergoing the same transformation from youthful and innocent fascination with the arts to the more serious world of politics. I could relate to their grievances and challenges to entering political life.

Why does the Cold War stigma of socialism hold such sway, even after decades of rising economic inequality, weakening labor rights, workplace burnout, declining levels of real pay and affordable housing, and skyrocketing carbon emissions? Neoliberalism has reinforced the stigma. It has scapegoated socialism as its fallible Other. Mainstream news outlets rarely offer socialists self-definition and provide little information about socialism. In the popular mind, socialism is intuitively framed as simply the opposite of capitalism. Why is it not recognized as a corrective to market fundamentalism aiming to create a more egalitarian, democratic, and sustainable world? Imagining only negative alternatives to the status quo is an intended outcome of othering. Stigmas are distorting, and that should make social scientists suspicious.

A colleague of mine proposed removing the word "socialism" from the title of this book to broaden its appeal. "How about 'New York Activists'?" they asked. Such small acts are more significant than one might think. I use the term "socialism" because it is the proper word. Democratic socialism is a force in New York politics, and socialist ideology is an important aspect of NYC-DSA. The organization's integration of ideology and organization represents a distinct alternative to mainstream approaches to political organizing and participation that separate these practices from relations of power and political outcomes. To ignore socialism, moreover, would be an undemocratic reduction of the diversity of ideological perspectives and exclude the entire historical project of the labor movement. It is therefore important for analytical and ethical reasons to reckon with the stigma of socialism, repair the damage, and counter the ongoing symbolic violence against the Left, including the climate Left, inflicted by Cold War propaganda, neoliberalism, and fascism.[26]

Setting Up the Investigation

How can we know what the democratic socialist movement is really about? How does NYC-DSA engage thousands of New Yorkers in organizing for climate legislation? Are the methods new? What are the lessons for future political struggles over climate?

There is valuable theory to help answer these questions, but readers wanting to get straight into the world of the movement can skip to chapter 3. However, the following introduction to the academic field in which this book is grounded should be accessible and help provide a deeper understanding of the movement.

Even if you do not know the name of the field—political sociology—you can surely relate to the motivating idea: Power relations in society shape political processes and have cultural dimensions. Class, race, and gender influence how citizens can express their concerns and gain political influence. If we insert climate into this equation, we have a brief diagnosis of much thinking about the issue over the past decade. Are you asking why climate activists are repressed and vulnerable populations neglected, while the fossil fuel industry has outsized power? If so, you are asking questions of political sociology.

This book adopts a political sociological perspective on the democratic socialist revival, seeing it as an attempt at mobilizing an alternative to the dominant forces of neoliberal capitalism and fascism. The book begins by reviewing the academic literature on the international emergence of socialist movement parties, interpreting them as responses to the Great Recession and the cleavage between political institutions and the working class. The literature further suggests that movement parties emerged as alternatives to the dominant organizational culture of political parties. This requires integration of perspectives from scholarship on movements and parties. This is also relevant because the literature on climate struggles is dominated by a movement-centric perspective, and the media reinforce this perspective by focusing on the visual spectacle of contentious protest.

The history of movement parties is thus part of the broader history of democracy in neoliberal capitalism, but the socialist movement in the US is also part of the history of US grassroots democracy and civil society organizing. A formative moment in this history was the debate between Walter Lippmann and John Dewey in the 1920s. Lippmann argued that politics in modern society is best placed in the hands of elected officials. Dewey found Lippmann's faith in elites undemocratic and argued that without an active citizenry, society would become an oligarchy. His perspective was embraced in the 1930s by the organizer and author Saul Alinsky, who pioneered the principles of community organizing that are now standard in labor-union organizing. Alinsky's work inspired the civil rights movement and was later popularized by the Barack Obama campaign in 2008. It was through the Obama campaign that Alinsky's approach was adopted by NYC-DSA. In the book *Blessed Are the Organized*, Jeffrey Stout draws on Alinsky's ideas to argue that grassroots organizing is essential to a just society. Grassroots organizing is especially relevant during local climate hazards, Stout argues, because it enables the community to hold governments and corporations accountable. In addition to the Alinsky tradition, NYC-DSA also draws on labor organizing methods and specifically the work of Jane McAlevey, who has done workshops for the organization.[27]

To explain political organizing in contemporary society, this book incorporates a broader sociological perspective on the changing relationship between place, body, and power. In particular, I revive Anthony Giddens's argument about reterritorialization for the analysis of political culture in New York City and develop this argument with media and communication theory. NYC-DSA reclaims embodied participation and communication for grassroots autonomy in the context of powerful neoliberal media, the erosion of local communities, and the rise of digital individualism.

Methodologically, the book is based on ethnography and research in the archives of news media, government, and DSA's various digital channels, including its national website and the Google Drive folders of relevant working groups.

Social scientists value ethnography for its capacity to bring the researcher into closer contact with social life. This is especially relevant to areas of social life that are controversial and poorly represented by the media. When I told colleagues about my fieldwork, some responded with reservation, indirectly saying, "This movement is just an echo of past leftist revivals. These are just young people in the city searching for identity and belonging. Do they not know that socialism is dead?" The language in most responses I have encountered relies on externally defined images. Understanding the self-definition of movement participants can be more than an act of kindness. It can provide insight into how citizens define and struggle for social justice. Ethnography is also essential to understanding organizations, especially urban grassroots organizations that are defined more by oral culture and informal networks than are other types of organizations.

I conducted inductive ethnography, approaching informants with a general question rather than with a precise question and theoretical framework to be tested. This approach is useful for discovering new phenomena and challenging existing theories. I initially asked my informants how NYC-DSA affords community, but I revised the question as I started to understand the informants better. The organization affords community, but what makes it distinctive is members' experience of meaningful political participation, significantly influenced by the particular political project and form of organization. My general interview question became, "What is distinctive about NYC-DSA's movement culture and its potential for influencing climate policy in New York State?" This created a focus on the organizational culture and movement party approach that began to develop in Brooklyn in 2017.

Because I focused on one electoral campaign and its immediate environment of Central Brooklyn, I could quickly gain a high level of ethnographic exposure. The concept of exposure refers to the amount of time that the researcher spends observing and interacting with the same person, organization, or place. The aim is to reach a level of saturation where interviews no longer reveal any new significant information. My in-person fieldwork lasted seventy-six days in spring 2022 and was followed by considerable digital research upon my return to Denmark up until fall 2024. Seventy-six days is long enough for the researcher to feel that they live in that place and to adopt a routine in everyday participation in a community. I quickly gained access to the organization because the barriers to entry in NYC-DSA are low for repeat canvassers. I gained a sense of important perspectives among a diverse group of people in the largest branch of NYC-DSA, observed how they acted in different situations, and conducted forty-four in-depth interviews. I also had follow-up conversations with key informants. Still, seventy-six days is not long enough to develop long-term relationships and experience longitudinal organizational change firsthand, but this is also not central to the book's objective. I conducted ethnography to discover new developments in political life and gain enough experience to meaningfully explain these developments. The digital fieldwork conducted remotely from Denmark involved a dozen more online interviews, follow-up communication with informants that included feedback on chapter drafts, and considerable digital archival research.

This is the first book written about NYC-DSA and the first academic account based on ethnography. It contributes foundational insights into the organization's history and culture, but it is not a complete biography of the organization. The book focuses on the organizational culture after 2016 and its potential for climate survival.

Such analysis is lacking in the magazine articles and book chapters that discuss DSA and democratic socialism in the United States. The popular 2020 book *Bigger than Bernie*, by the two *Jacobin* editors Meagan Day and Micah Uetricht, is a great introduction to the formative moment in 2017 but does not analyze the organizational culture. It was completed in 2019, before the electoral work matured. There is a new literature on socialism in the United States that focuses on its national agendas and public figures, which is similarly neglecting the organizational turn in NYC-DSA and the broader international political history. This literature includes the 2020 edited volume *We Own the Future: Democratic Socialism—American Style* (edited by Aronoff, Dreier, and Kazin) and Garry Dorrien's *American Democratic Socialism*. Kim Moody's *Breaking the Impasse* explores the Democratic Party's interaction with movements in the twentieth century and makes a case for independent organizing. Moody says little about organized life in the contemporary city, however. He describes DSA as "the organizational expression of this new socialist movement" but does not explore it.[28] Matthew Huber's *Climate Change as Class War* (2022) is a foundational contribution to Marxist theory of the climate crisis with a focus on structures, not agency, organization, and political processes. I hope to fill this lacuna.

My Journey

This book emerged from the circumstances of an overseas research field trip, the COVID-19 pandemic, and the climate crisis. I tell the story of my field experience for transparency and to stimulate reflexivity about how social scientists act on their experiences of this crisis. My experience is privileged. Many people have died from the pandemic and climate hazards. During this time, I lived comfortably and was able to write a book. The project began in 2021 when I was planning to conduct fieldwork in New York City for a book on how digital media are reshaping community. The plan was to interview high school students in two Brooklyn neighborhoods. The pandemic scuttled this plan. All schools were in lockdown when I arrived in New York City in March 2022, and physical distancing was practiced in most public places.[29]

My plan B was to do armchair research. But after two years of pandemic life, I had a strong urge to talk to people, and I much prefer to write empirically grounded books. I had landed in the affluent Brooklyn neighborhood of Park Slope and began exploring the area by bike. On one of my first trips to nearby Flatbush, I stumbled on volunteers petitioning to get a person named David Alexis on the ballot in the upcoming state senate race, and they talked about democratic socialism and climate change. This encounter triggered my interest. I had not experienced anything like the culture of this organization before. This was an unusually vibrant grassroots organization with a clear focus on issues that seemed relevant to large urban populations, including a wealth tax, labor and tenant rights, and climate change. The movement rhetoric caught my attention, and my interest grew when I learned more about the organization. The Alexis campaign was one of six climate-themed electoral campaigns that pressured leading politicians in state government to pass climate legislation developed by the grassroots in NYC-DSA. The participants were different from the image I had of socialists. They were not sectarian or intellectualizing. They had relevant arguments and practical solutions and were not preoccupied with socialist identity. The participants valued particular conditions of political participation in this organization, and I could intuitively relate to this culture because of my experience in the city's independent arts field. Some of the participants had come from the arts scenes in northwest Brooklyn, where I had done fieldwork from 2010 to 2013.

The encounter with DSA became the catalyst for a transformation in my thinking. For two decades, I had been writing about the history and sociology of music. This framing of my research was starting to feel constraining. I had become concerned about the growing problems of neoliberalism and climate change. It would have been easier for me to continue writing about music if it had a major role in the climate crisis, but this has not happened yet because climate change motivates people to focus on basic elements of survival. Climate change inverts Abraham Maslow's famous hierarchy of needs. My native, affluent Scandinavia focuses on the top of this hierarchy, namely, self-actualization, but our basic physiological needs are threatened, including food, water, and shelter. "A person who is lacking food, safety, love, and esteem would most probably hunger for food more strongly than for anything else," writes Maslow. He continues, "[if] the organism is then dominated by the physiological needs, all other needs may become simply non-existent or be pushed into the background." In NYC-DSA, I met several people who, like me, had transitioned from arts to politics, and for the same reasons. They had gone from working for arts organizations to political NGOs and devoting much free time to NYC-DSA.[30]

My response to NYC-DSA was also influenced by the experience of returning to the United States for the first time since 2013. Scandinavians are often enthusiastic when they go on vacation in the United States, but those who live there for a longer period of time often find themselves taken aback by the extent of poverty, gun violence, and Christian nationalism. This experience has led to books such as Any Partanen's *The Nordic Theory of Everything* (2016). Partanen describes her experience moving from Finland to New York after the Great Recession as "an extraordinarily harsh" form of travel backward in time.

She is surprised by the lack of knowledge of the Nordic welfare state model. This was before Trump's first term and before the pandemic, and the impacts of those events were tangible when I returned in 2022. Scandinavian media cover US presidential elections with overwhelming support for the Democratic Party's candidates, but I believe that many Scandinavians would be sympathetic to US democratic socialism if they lived in the United States.

My Journey in DSA

NYC-DSA is a diverse organization with eight neighborhood branches across a large urban area. I focused on the organization's networks in Central Brooklyn, the epicenter of "the new NYC-DSA." I canvassed regularly for an electoral campaign in this area, the David Alexis campaign, which led to the subsequent creation of a new branch, the Flatbush branch. I canvassed for a campaign because this is the most meaningful way to become involved in this organization. Canvassing is the universally shared activity of its active members and leaders. I talked to hundreds of voters and people in the campaign office on Flatbush Avenue. The task of my weekly canvasses was to persuade registered Democrats to vote for DSA's more left-leaning and climate-progressive candidate. I explored how voters feel about politics, their grievances, and how they responded to democratic socialism and its climate politics. The Alexis campaign had strategic importance, aiming to expand DSA southward in Brooklyn and build momentum for Green New Deal legislation. It directly challenged Kevin Parker, the chair of the Energy Committee in the state senate.[31]

Electoral campaigns in NYC-DSA are not limited to the individual branches, and through the Alexis campaign, I developed relations with the organization's citywide networks. Once the electoral campaigns begin, they take on a life of their own, recruiting volunteers from all over the organization and the city. In this case, the pioneering Brooklyn Electoral Working Group was in charge. However, the organization's citywide climate working group, the Ecosocialist Working Group, had played a key role in developing the platform for the six climate-themed campaigns in 2022 that electoralized the "Public Power" movement campaign. That campaign had been building for three years but had thus far failed to win. The electoral campaigns thus represent a key movement party process of electoralizing a movement's political project. The managers of the Alexis campaign and several of the canvassers had been deeply involved with the Ecosocialist Working Group for years.

Through the Alexis campaign, I also developed relations with citywide leaders and explored the national context of democratic socialism and the Green New Deal through research on movement literature, Sanders's autobiographies, leftist magazines, news media, and the archives of NYC-DSA. The archives of the state and federal governments provided details on socialist climate legislation.

NYC-DSA is a multitendency organization, and my focus on electoral work positioned me in the territory of the dominant factions with a democratic socialist position. I focused on the organization's electoral work because it is central to the organization's history after 2016 and the theme of this book. A discussion of perspectives among the small number of communists and anarchists is beyond the scope of this book. These groups think that Sanders's social democratic position is not radical enough and will lead to conformity with the status quo. I hope that they will nonetheless find this book useful.[32]

Organization of the Chapters

The book progresses from political history to fine-grained ethnography and organizational analysis.

The main steps in the inquiry are as follows: (1) a history and conceptualization of democratic socialist movement parties after the Great Recession and their particular articulation in the United States, (2) an ethnography of one high-stakes electoral campaign in Brooklyn and its relationship to the local Democratic Party organization and the long-term movement project of creating Green New Deal climate legislation, (3) an organizational analysis of NYC-DSA and its party dimension to explain in more detail how the organization is building capacity, and (4) analysis of how climate activism developed in this organizational context and how it led to legislation. A key point of my overall narrative is that the analysis builds from the microworld of one campaign—a high-stakes campaign for the organization's expansion and climate efforts—to the macrolevel of the organization as a whole.

Chapter 1, "The Climate Movement Needs Politics!," addresses the climate movement's skepticism about institutional politics. I argue from a social psychological perspective that this relationship is complicated by a psychology of neglect that derives from decades of government inaction, resulting in a narrow focus on protest and direct action. I subsequently broaden the perspective to argue that the climate movement's skepticism is amplified by the more general vicious cycle of declining political participation and the crisis of democracy.

Chapter 2, "The Sociology of Movement Parties," takes stock of existing knowledge in political sociology. I argue that political sociology has gained urgent importance in the crises of inequality, democracy, and climate. This subfield is concerned with political organization and power relations in society. It helps explain the role of movements in times of crisis. I situate this study as part of a Marxist revival in the social sciences. The chapter concludes with the movement party scholarship of Donatella della Porta and colleagues to discuss the dynamics of movement parties in late neoliberal capitalism. It also highlights the potential for looking beyond the movement-centric perspective that has dominated movement scholarship for decades.

Chapter 3, "The Political Environment After the Great Recession," understands the rise of the DSA in the late 2010s as the culmination of a decade of opposition to neoliberal capitalism. The chapter explores how NYC-DSA members responded to major political events and how the organization has framed itself in recent political history. Drawing from social movement theory, the chapter considers how these events were interpreted as crises and opportunities. The chapter shows that this urban movement was shocked by Trump's election, like the broader anti-Trump movement, but that it interpreted the situation differently and organized differently. It adopted a universalist political program focused on inequality and developed into a movement party.

Chapter 4, "Anticommunism Kills," examines the socialist movement's struggle for self-definition against powerful externally defined images. It shows that powerful proponents of neoliberalism and fascism are using anticommunist propaganda to frame the revival negatively. The chapter then asks, What does the term "socialism" actually mean to participants in the movement? It argues that the revival is primarily a response to extreme inequality and understands socialism in terms of social justice. The movement's focus is on concrete issues and practical solutions rather than theoretical debate. The chapter concludes by acknowledging that the international revival has had limited success in national elections in recent years but that it remains a distinct component of the political landscape and that some of its ideas have become folklore among younger generations. I illustrate the latter point in a discussion of the work of Sally Rooney and Jenny Odell.

Chapter 5, "A Departure from the Democratic Party in Brooklyn," situates NYC-DSA in the political field in Brooklyn and introduces the 2022 David Alexis campaign in Flatbush in this context. It argues that NYC-DSA is a departure from the Democratic Party's more traditional culture and a moral alternative to the party's "machine politics." The machine tradition of corrupt party bosses, patronage, and exploitation of immigrant populations originates in the Tammany Hall organization. Tammany dominated the party in New York State for about a century from the 1850s onward. NYC-DSA runs candidates on the Democratic ballot but is otherwise organizationally independent from the Democratic Party, and it challenges the party's centrism and moral values.

These dynamics are illustrated by the 2022 Alexis campaign, which was a key part of the electoralization of NYC-DSA's "Public Power" campaign. Alexis ran against the nineteen-year incumbent, Kevin Parker, who was supported by the party establishment, had personal ties to powerful local institutions, and took donations from fossil fuel companies. By contrast, Alexis was a working-class candidate struggling to gain a foothold in neighborhood institutions dominated by the elites, thus illustrating NYC-DSA's challenges in the political field. Alexis was so poor that he had to work full-time as a rideshare driver throughout most of the campaign, obviously resulting in suboptimal performance. A further complication was the advent of a spoiler candidate. Still, Alexis got 7,047 votes against Parker's 8,543, and the campaign pressured Parker to support the New York State Build Public Renewables Act that NYC-DSA had championed since 2019.

Chapter 6, "The Alexis Campaigners in Flatbush," explores the roles and identities of the people who worked to elect Alexis in 2022. The chapter begins with an ethnography of the campaign office before exploring the perspectives of the campaign management and canvassers.

The different roles in the campaign organization are introduced and illustrated with a few examples. The chapter explains what participants did, how they attached meaning to it, and how their work related to their political identities. It considers the individual's social background, education, and other experiences that shaped their relationship to the organization and to politics more broadly.

These insights into the Alexis campaign lay the groundwork for understanding the broader organizational culture and leadership in NYC-DSA as a whole. Chapter 7, "Organization and Leadership in NYC-DSA," provides an account of this organization's overall transformation in the late 2010s, focusing on developments in organizational structures and leadership practices. The chapter highlights continuities and change between the "old" NYC-DSA and the "new" organization that emerged after 2016. The new NYC-DSA is focused on organizing campaigns and influencing legislation. After 2016, organizational changes were made to strengthen grassroots democracy, and new caucuses and thematic groups were created, including the Ecosocialist Working Group. Moreover, the organization adopted an established movement leadership approach at the individual campaign level and at the overall organizational level. The chapter concludes by exploring the individual perspectives of three leaders to nuance the structural perspective.

Chapter 8, "The Party Dimension," expands the organizational analysis to the party dimension. It begins with NYC-DSA's overall goal of democratizing the political process in response to the situation in the Democratic Party. It then shows how the organization's party identity and position in the party system have evolved from movement-based electoral campaigns. There are growing expectations for clarifying the party dimension. The chapter also reveals limitations of the distributed leadership approach. It concludes that the potential of NYC-DSA in the current two-party system is to contribute to a transformation of the Democratic Party, while retaining political independence through the movement sphere. NYC-DSA is a movement-based organization that seeks to influence institutional politics through movement pressure campaigns and by building representation in state government.

Chapter 9, "The Road to Green New Deal Legislation," shows how the Ecosocialist Working Group in NYC-DSA developed and won transformative climate legislation for New York State, namely, the New York

State Build Public Renewables Act of 2023. The chapter begins by introducing the history of the socialist Green New Deal resolution, introduced to Congress in 2019. This proposal represents a change in the country's climate movement and influenced the Ecosocialist Working Group. The chapter then examines the group as a distinct organizational environment in NYC-DSA with ties to professionals in the climate field. The group led a statewide coalition with climate organizations and progressive Democrats. It used a range of tactics and worked across movement and institutional contexts. It translated movement narratives into institutional genres and electoralized the movement campaign. The movement's collaboration with its elected officials played an important role, demonstrating that the success of a movement party does not come from the organizational design alone but also from the practice of collaboration and trust across organizational contexts.

The first of two concluding chapters, "Conclusion 1: Status of the Democratic Road to Climate Survival," evaluates the achievements of NYC-DSA in the context of the evolving national political environment. It begins by acknowledging that the organization has not fundamentally changed the political situation and struggles to build long-term organizational capacity. I caution against a simplistic conclusion of failure or success and highlight the significance of seemingly small achievements, arguing that small achievements and opportunities are essential to growing the movement and achieving larger goals in the future. The democratic socialist movement has initiated an insurgence in neoliberal capitalism and shaped how many people think about climate change. NYC-DSA has institutional achievements, and its greatest potential for the democratic road to climate survival lies in the alternative to the Democratic Party organization, specifically its demand for a socially just energy transition and democratic, movement-based political organizing. From this perspective, I show how small changes could have turned some of NYC-DSA's losses into victories and why the principles of collaboration across grassroots and electoral politics are relevant in the "sticky" climate crisis.

The final chapter, "Conclusion 2: Scholarship for Survival," suggests how scholars of climate politics can do more to help us survive climate change. It first confronts two major forces in political culture today, namely, media and corporate capitalism. Such systemic aspects are important to recognize, but scholars can learn from political grassroots organizers to overcome the apathy that often results from diagnostic analysis. Scholars of climate politics can learn from activists and political organizers to increase the direct and practical relevance of their work to climate survival. They can also collaborate more with progressive climate forces in politics. I argue that scholars can advance the cause by further integrating the disciplinary perspectives emerging from this book, namely, movement leadership, political analysis, and ethnography.

1

The Climate Movement Needs Politics!

I began this book by arguing that the democratic road to climate survival is complicated, but giving up on politics would be worse than the alternative. It would lead to more authoritarianism and disaster capitalism. In this chapter, I argue that the first step toward popular democratic solutions is overcoming the negation of institutional politics. Drawing on a social psychological perspective, the chapter shows that decades of government inaction on climate have led to the climate movement's rise but also its growing alienation from institutional politics. A growing feeling of government neglect has become an obstacle to democratic solutions, accelerating the vicious cycle of declining political participation. Another problem arising from government neglect is the climate movement's perception that there is no time for long-term political and organizational processes, only direct action and protests. I conclude the chapter with an outline of forces in broader political life that contribute to the vicious cycle of declining political participation. The next chapters will show how New York socialists are trying to change this situation.

The Climate Movement's Relationship to Politics

One of the most influential political theorists of the past fifty years, Jürgen Habermas, argued that social movements have the unique role of sounding the alarm bell when institutions fail to act in times of crisis. Movements have this function because they provide citizens with a direct way of organizing and communicating their concerns. They have a unique role in political life, moreover, by moving ideas from the outermost periphery into the public sphere and the political system. A problem not anticipated by Habermas is the sticky crisis that evolves when the alarm bell has been rung for decades and society is far from making the changes necessary to prevent climate collapse. The psyche of public opposition grows increasingly desperate and exhausted amid this extraordinary crisis. There is an evolving tension between a sense of endless government neglect and the mounting urgency of climate change. These social psychological aspects have implications for perceptions of political participation and democracy.[1]

Neglect

Decades of government inaction can be understood as a form of institutional neglect of citizens' safety that registers in public consciousness. It affects how citizens view political institutions. I draw inspiration from the psychology of neglect that has developed in research on child development. Psychologists are yet to develop a theory of government climate neglect, but there is ample evidence of this phenomenon.

A revealing historical entry point is the climate movement's changing response to the biggest international events of climate politics, the biannual Conference of the Parties (COP) organized by the United Nations. The climate movement has been disappointed by every conference and for decades tried to mobilize hope for the next conference until gradually abandoning it. Today, many prominent climate activists have lost confidence in this so-called global climate regime, countries in low-latitude countries have become highly skeptical, and the leaders of China, United States, and Germany skipped the meeting in Azerbaijan in 2024. This was the meeting where the host described oil as a "gift of God."[2]

The first sign of a shift from short-term disappointment to a sense of long-term neglect can be found in the aftermath of COP 15, in 2009. The movement came with high expectations, boosted by the presence of Barack Obama, who had just been elected president on a progressive platform. It turned out, however, that political leaders were primarily interested in planning the economic recovery from the Great Recession. In a meeting with a small group of leaders hours before the conference ended, Obama said he would not stay beyond a few hours, "because all of us obviously have extraordinarily important other business to attend to." The conference ended in a nonbinding agreement. In *This Changes Everything: Capitalism vs. Climate* (2014), Naomi Klein describes an acute and painful realization that our "leaders are not looking after us. . . . We are not cared for at the level of our very survival." Extinction Rebellion (XR) has repeatedly called out government failure and demanded that the government declares a climate emergency. Greta Thunberg has called out politicians and adults in general numerous times for failing to protect the future of younger generations.[3]

The psychology of neglect is also evident in climate litigation against governments. Claims about government negligence have been made since the early 2010s against governors in the United States for repressing conversations about climate despite this information having consequences for citizens. Attempts have been made, moreover, to argue that states should not be immune to lawsuits about failing to prepare for climate hazards. Internationally, a 2015 court decision in the Netherlands forced the government to reduce greenhouse gas emissions. This victory inspired the creation of the Climate Litigation Network, which has since helped organizations around the world in about eighty climate litigation cases against governments. Some legal scholars argue that a human rights approach could be more successful than a negligence approach, but they are not denying the neglect.[4]

The psychology of child neglect focuses on children's relationship with their immediate family and community, referred to as the environments of primary and secondary socialization, respectively. However, some insights are relevant to the national environment of government and media, which can be understood as the tertiary environment of socialization. They include the lack of care for physical safety having severe consequences for a person's growth and well-being, such as anxiety disorders and complicated social relationships, including social withdrawal.[5] The comparison with child neglect also helps explain that citizens are not experiencing neglect of small problems. Climate is a big and existential societal issue. One symptom of this psychology of institutional neglect is the resentment that surfaces in aggressive disruptions of political fundraiser events and annual general meetings of corporations, for instance, with climate protesters screaming and shouting. Another symptom is the emphasis on ad hoc protesting focused on the here and now, with the underlying assumption that there is not time to build democratic organizations. "Alarmed about climate change? Join our protest against a big carbon polluter!" The short-term focus on disruptive protest can partly be explained by limited organizational resources. Movement organizations with few resources can have some impact through disruptive protests, but there are limitations to this strategy.[6]

The feeling of neglect is weaker among the broader public than in the climate movement. This is suggested by the fact that the majority of Americans have become concerned by climate change but believe that it is still possible to turn the situation around. This momentum cannot be leveraged through short-term protests and the escalation of tactics alone.[7]

Escalation

The feeling of institutional neglect stimulates the escalation of tactics. It stimulates anger, desperation, impatience, and tunnel vision. The escalation can be identified in the confrontational and spectacular forms of activism performed by the radical flank of the protest movement: vandalism, roadblocks, and sabotage. The leading organizations include Extinction Rebellion (2018–present), Letzte Generation (Last Generation; 2021–present), Les Soulèvements de la Terre (The Earth Uprisings; 2021–present), Just Stop Oil (2022–present), and Climate Defiance (2023–present). In the US, the radical flank has grown from disappointment with the Biden administration's Inflation Reduction Act of 2022, thus further illustrating the dynamics of neglect.

A distinctive feature of the radical flank is the goal of shocking the public and getting arrested. These "shockers" are sounding the alarm bell more loudly, although they are nonviolent and do not threaten the safety of citizens. Their tactics require a more secretive style of organization, including anonymous digital identities and encrypted communication, to protect participants against legal prosecution. This form of communication is prohibitive to movement solidarity and aspects of democratic culture, such as public transparency and deliberation. The culture of radical protest also involves a high level of organizational instability, with protesters quickly burning out and being imprisoned. At one point, XR declared that it was shifting emphasis from disruptive to constructive tactics, but protests remain central to this organization. In 2024, the Austrian branch of Letzte Generation dissolved because it could no longer see a meaningful way forward with its tactic of blocking road traffic. The group faced major obstacles from all sides, citing imprisonment, fines, and death threats. The group had also been criticized by politicians, including representatives of the Green Party who argued that its tactics discouraged public interest in the climate struggle.[8]

Movement-centrism

How are scholars responding to the climate protest movement? The movement literature has provided important insights, but it suffers from decades of movement-centric thinking. Dana Fisher has made foundational contributions to climate movement history but refrains from analysis of broader dynamics in the political environment. Her work does not consider the development of the climate Left and new urban political movements, for instance. This is relevant because the protest movement is evolving dynamically in relation to this broader political environment and because the climate Left is channeling protest into institutional politics. Moreover, Fisher is one of the advocates for the idea that mass mobilization will only happen when climate hazards get more severe. While I agree that a greater sense of risk creates potential for political change, constituting what sociologists call an opportunity structure, I do not think that we can afford to wait for this, and it will not happen if we do not build more democratic and socially just political institutions. We therefore need to place central emphasis on the politicizing and organizational processes that can leverage the potential for climate survival.

Andreas Malm and the Zetkin Collective have introduced an engaging Marxist interpretation that calls out fossil fuel capitalism, but Malm focuses exclusively on building pressure on institutions from the outside. His work has thus far not included the perspective of building popular democracy. *How to Blow Up a Pipeline* makes a case for confrontational direct action, including sabotage, blockages, and property destruction, arguing on historical grounds that such forms of action have played a decisive role in winning some of the most important fights for social justice throughout modern history. I agree that direct action has played an important role but crucially in combination with broader coalitions that include political parties. In short, scholarship on climate protest could usefully reframe its perspective, taking inspiration from the critique of movement-centric thinking in political sociology.[9]

A similar lack of interest in building popular democracy can be found in the so-called coercive turn in movement studies. A small network of radical theorists is defending the normative legitimacy of coercive civil disobedience. The liberal view held by John Rawls is that the civil dimension of protest is undermined if it is coercive: blockages inhibit the democratic process. Protest can warn, but it should not threaten. The coercive turn challenges this thinking in political liberalism, claiming that it represents the idealized view of a privileged elite. I agree, but we need to look elsewhere for political analysis and organizing.[10]

An Organization Based on Protest Events

To give a more concrete sense of how the psychology of institutional neglect shapes the climate movement, I now turn to the everyday life of XR in Denmark. In spring 2023, I participated in protests organized by the organization's academic arm, Scientist Rebellion (SR). SR Copenhagen is led by a small group of international researchers and constitutes a kind of "protest cell" in an international network. All of its communications are in English. This helps SR reach international networks, but it creates a distance from the local public. Typical of the radical flank, SR Copenhagen is focused on disruptive protests and direct actions that lead to arrests, with much coordination happening in the encrypted app Signal. It has occupied public offices and blocked traffic. The cell has a pool of about twenty repeat activists and an inner circle of coordinators led by Fernando Racimo, a professor of evolutionary genetics at the University of Copenhagen. The coordinators recruit people in their network to participate in individual protest events. The inner circle sets an example by getting arrested, but they emphasize that this is not required for participation. I admire SR Copenhagen for being one of the few organizations in my hometown that have consistently organized climate protests, and I admire Racimo for taking courageous action. He was one of the few professors who publicly supported the local student protests for peace in Palestine in 2024, and he has fearlessly spoken up on issues such as academic freedom. While I fully acknowledge the relevance of climate protest, I think it is important to think critically about the limitations of SR and similar protest organizations. In particular, their exclusive focus on protest without broad-based democratic organizing leaves a void that can only be bridged by other organizational frameworks. I explore this view further in the following ethnographic excursion.

Copenhagen, Denmark, May 10, 2023, noontime. I am entering the headquarters of the Danish Agriculture and Food Council along with seven other university researchers and a choir organized by XR. This industry association is one of the most powerful lobbying organizations in the country, and it has become the target of repeated protests from climate movement organizations. Last year, XR activists glued themselves to the pavement outside the building.

As we approach, I do not feel prepared and ask myself how I got into this situation so quickly. Some participants have no other training than a briefing a few hours before the event. Yesterday, we received literature on the agricultural industry's climate footprint. Entering the building without an appointment is surprisingly easy, and the situation is at first calm in the majestic hall of this neoclassical building. The architecture is a visual expression of the country's nineteenth-century elite. It is located in the capital city, right next to the offices of other national industry associations and international corporations, the city hall, and the central rail station. I find it striking that the building has remained unchanged for a hundred years, with no sign of the owners' environmental footprint. But today, a transnational twenty-first-century climate movement is intervening in this space and reframing it through social media platforms from a posthuman anarchist perspective.

The singers spread across the first four levels of the staircase, while the researchers take a seat in the middle of the hall on the ground floor (we brought folding chairs). We are thus claiming a collective spatial presence. "Humans. King of the Ashes. Our shit is muddying the waters . . ." The dystopian lyrics are sung slowly and gently, with long notes resonating in the hall, which has the acoustics of an old church. This serene performance does not sound confrontational at all, but it frames industrial animal farming as the agent of catastrophe. When the song has ended, the singers let their veils levitate slowly down to the ground, and they unfold a large banner with the words "The Last Pigs." Racimo stands in front of the researchers and delivers the opening speech:

Together with more than one thousand scientists around the world and Extinction Rebellion, today we are engaging in civil disobedience to demand immediate climate action to industries and governments. We are here at the Danish Agriculture and Food Association, a lobbying body representing the farming industry of Denmark, which is currently responsible for 30 percent of Danish emissions. The science is clear: the IPCC says we must stop emitting greenhouse gases now. The association claims that they will show the world an economic and sustainable way to a climate-neutral food production, yet they have absolutely no strategy for transitioning and the emissions have not decreased.

Next up is a sociologist of food: "We have failed to deliver change, even when citizen assemblies show clear support for a transformation of the food system."

The speeches are interrupted by a security guard. He tells us that he has called the police and is hateful. Climate activists are the red rag to this bull. He is harassing a woman trying to leave before the police comes. His expression of brute force is intended to make us uncomfortable, and it works. Four police cars arrive with sirens in a full-on emergency and shut down the building. All because of a tiny, peaceful protest! The protest did not even disrupt life in the building. Some office workers stopped to enjoy the choir performance. "Did they do any harm?" asks a police officer. "No," the guard replies. We are allowed to leave, but only after being photographed by the police.

The event illustrates XR's dramaturgic framing of the climate struggle: Protesters identify an organization responsible for massive carbon emissions, frame it as a villain, and occupy its space to make moral statements. During the event, I felt that we were engaging in a heroic act of trying to stop climate destruction. It seemed right to ask the public to stop and think. I felt contempt toward the guard and the police for their hostile behavior toward climate activists. But then nothing happened. The organizers simply invited us to the next protest event. They offered a debrief but did not involve participants in a decision-making process about future activities and goals.

The Politicization of the Climate Movement

Another barrier to the democratic road is the climate movement's promotion of apolitical understandings of climate change. The movement has struggled for decades to persuade the public that climate change is not a lie, so bracketing politics has been a way of claiming objectivity. It has also adopted an apolitical position to build support across political divides.

The idea that apolitical understandings of climate change are objective and positive originates in the most powerful institutions in society. In the 1980s, governments made conscious efforts, in alliance with the corporate world, to define climate change narrowly as an environmental problem. Governments adopted a technocratic perspective that decontextualizes climate from society. Governments and private foundations still support this "scientized" understanding of climate change that removes moral and political considerations from the equation. They create the conditions of a discourse, through research funding and communication, that frames climate change as the outcome of an abstract and universal "humanity" or "consumer needs."

Climate sociology evolved as a corrective to this apolitical thinking. It expands on the tradition of environmental sociology since the 1970s that critiques the decoupling of economy and ecology. Climate sociologists argue that climate change is rooted in socially structured forms of production and consumption. For instance, consumer behaviors are embedded in the value system of capitalism, and market forces stimulate the continued growth of unsustainable industries. The market logic explains the absurd situation in which fossil fuel and petrochemical industries try to keep society in a carbon lock-in and maintain their social license to operate by presenting themselves as enablers of the transition to sustainable energy. Climate sociologists such as Robert Brulle have pioneered investigations into the public relations practices of these industries, including the development of a politically influential anticlimate movement. This movement is a coalition of fossil fuel companies, lobbying groups, conservative think tanks, and powerful individuals such as Mitch McConnell and Rupert Murdoch. It has for decades promoted a positive view of fossil fuels, delayed political opposition, and redirected responsibility to the consumer. In 2015, research into industry documents found evidence of a decades-long and still-ongoing campaign by "the world's major fossil fuel companies and their allies to spread climate misinformation and block climate action." In response to changes in public opinion, carbon-intensive industries now acknowledge climate change and instead promote narratives of climate delay, downplaying the urgency and justifying inaction.[11]

There are signs of a political reckoning in the climate movement in the early 2020s. This politicization is inspired by the new climate Left and its popular intellectuals, including Naomi Klein, Kate Raworth, Jason Nickel, and Andreas Malm. Thunberg's speech at Davos in 2019 signaled a turning point. Up until then, she had appealed to the moral conscience of political leaders, but she now began to focus on their failure, adopting a revolutionary rhetoric. Thunberg is no longer simply saying, "Listen to the science." After touring the halls of power around the world and urging the elites to listen, she has developed a political criticism of those elites and participates in direct actions. In 2022, Thunberg said that she did not attend the UN's climate summit because it serves elite interests. A few months later, she adopted a leftist perspective on the World Economic Forum: "We are right now in Davos, where basically the people who are mostly fueling the destruction of the planet, the people who are at the very core of the climate crisis, the people who are investing in fossil fuels, and yet somehow these are the people that we seem to rely on solving our problems when they have proven time and time again that they are not prioritizing that; they are prioritizing self-greed, corporate greed, and short-term economic profits above people and above the planet."[12] This narrative is now widespread across the youth climate movement, which is no longer simply telling us that climate change is a fatal risk; it is also interpreting it as a political crisis. I learned just how normalized leftist climate narratives had become in the youth movement from conversations with Fridays for Future protesters in New York City in 2022. At one of the Friday protests, an activist held a poster saying, "Cause of death: Capitalism."

The politicization is also happening in parts of XR. The organization's leadership in the United Kingdom has argued that a broader political perspective is a natural evolution from pressure campaigns. The leadership issued a statement on New Year's Day in 2023:

> When XR burst onto the scene four years ago, few could have imagined the seismic shift it would bring about in the climate movement, the climate conversation, and the world at large.
>
> But despite the blaring alarm on the climate and ecological emergency ringing loud and clear, very little has changed. Emissions continue to rise and our planet is dying at an accelerated rate.
>
> The root causes? A financial system prioritising profits over life, a media failing to inform the public and hold power to account, and a reckless government entrenched in corruption and suppressing the right to protest injustice.
>
> As we ring in the new year, we make a controversial resolution to temporarily shift away from public disruption as a primary tactic. We recognise and celebrate the power of disruption to raise the alarm and believe that constantly evolving tactics is a necessary approach. What's needed now most is to disrupt the abuse of power and imbalance, to bring about a transition to a fair society that works together to end the fossil fuel era. Our politicians, addicted to greed and bloated on profits won't do it without pressure.[13]

The announcement signals an interest in expanding the organization's perspective from protest to movement building. It also signals a broader perspective on political life. I welcome this development and seek to push the perspective further by addressing the gap between movements and institutions. There are good reasons to feel hopeless about the political system, but negating institutional politics altogether would amplify a vicious cycle in political life. Protests are still relevant, but protests alone cannot build the organizations and politics that we need.

General Barriers to Political Participation

I conclude this chapter by recognizing that the psychology of neglect is amplified by a more general vicious cycle of declining political participation in society. This vicious cycle is mainly created by the following four forces.

Polarization

Political scientists have documented growing polarization since the 1960s. There is increasing popular support for extremist political views relative to the support for centrist or moderate views. Negative partisanship is on the rise, with more people voting for one party only because of negative feelings for the other party. Polarization also refers to feelings of resentment and distrust against people identifying with another party. Cable news stations have contributed to polarization, enabled by political deregulation, with the Federal Communications Commission abolishing the fairness doctrine in 1987. Since then, broadcasters no longer have to present political news, and if they do, they are not required to present different political perspectives. A cable news culture has emerged in which anchors are using vicious language, mockery, and insults. The political scientist Eitan Hersh finds that cable news has turned politics into a form of entertainment, amplifying drama and focusing on celebrities and scandals. Presidential elections have become gladiator fights.[14]

I found little evidence of polarization in NYC-DSA, although there were feelings of contempt and disgust for the Democratic Party. These feelings are common in the moral commitments of social movements, as James Jasper has pointed out. They indirectly position other groups as outsiders and thus create social disconnection and division. They can spiral into a process of treating others as subhuman, but this is far from happening in NYC-DSA. The organizational culture is too dominated by positive and constructive emotions, built into its core activities of community organizing and canvassing, for this to happen. Its emphatic commitments to democratic values and egalitarianism further prevent such negative tendencies.[15]

New York socialists are not driven by hate or waging a culture war. Their indignation about the far Right and the Democratic Party's establishment is a motivation, but it does not consume them. The meetings I attended in the Alexis campaign and the Central Brooklyn branch were not filled with negative emotions or negative statements about others. Although a few Democratic leaders were sometimes described as "villains," this critique targeted their political actions and was articulated with a sense of republican-democratic values of respect for democracy.

The Alexis campaign office hosted a grassroots culture driven by a vision for a just and sustainable future, not a hate crowd. Also, the socialists did not promote conspiracy theories. They were subjected to them.

The socialists were subjected to negative polarization from the outside, specifically in the form of anticommunist bullying. I agree with the political scientists Jacob Hacker and Paul Pierson that polarization in the US is asymmetric because it is more deeply embedded on the Right than on the Left. I am therefore also skeptical of the horseshoe theory, which says that the Right and the Left are similar at the extremes. The "both sides" gesture implies that they are equal, but the far Right is the main aggressor of polarization, and it engages in antidemocratic behaviors. Trump has stimulated the rise of hate crimes and celebrated the insurrection in 2021. The Right has also gained dominance in political and media institutions in neoliberal capitalism. The Left is prodemocratic, and it is much smaller. One element of polarization that I did find in NYC-DSA is the element of demographic homogeneity, although the evidence for this is somewhat ambiguous. Polarization scholars talk about "social sorting" in politics, following the trend of citizens increasingly living in neighborhoods with members of their own social group and party. NYC-DSA is concentrated in certain neighborhoods and somewhat homogeneous with regard to age and class, but it has become increasingly diverse with regard to race and ethnicity.[16]

Gerontocracy

The overrepresentation of old white men is another democratic problem. People under age thirty represent half of the world's population but only 2 percent of legislators. In the United States, five of the most powerful politicians in the early 2020s—Joe Biden, Donald Trump, Nancy Pelosi, Chuck Schumer, and Mitch McConnell—were white, over seventy, and wealthy. The average citizen was approximately twenty years younger than the average member of the House of Representatives. This gerontocratic patriarchy is supported by institutional structures that limit access for younger and more diverse people and normalize prejudice against them.[17]

In the field, I observed how the word "young" was frequently used as a pejorative by political opponents. I found a new version of the Vietnam protest phrase "old enough to fight, old enough to vote" among teenagers protesting at City Hall with Fridays for Future. One of them, Helen Manchini, described the paradox of being born into a world of climate change but excluded from political influence because of age. She related strongly to Greta Thunberg, "[who] was always talking about how all the adults failed us," thus connecting the exclusion of young people from the political process with climate neglect. Manchini continued, "When you're a kid, no matter what your situation is, you feel sort of out of power. A lot of adults have so much power over you, and you can't vote, and there's this doom that is put onto you just by being born. In some way, skipping school, leaving early, and doing something that you created without the help of any adults is just getting that power back. You don't have actual power in democracy, but it's just action that no one can take away from you. They can't arrest you. They can't make you stop." The phrase from the antiwar movement could be updated for the climate crisis to "old enough to know climate change, old enough to vote."[18]

The problem of gerontocracy is widely acknowledged in the anti-Trump movement. Some of the organizations that emerged to rebuild democracy after Trump's election recruit young candidates. Run for Something, for instance, recruits and trains candidates under forty. It supports candidates across the spectrum on the Left, especially progressives, but it has also supported DSA candidates. The organization is co-led by Amanda Litman, who worked on the communications team for Hillary Clinton's 2016 presidential campaign. I asked Litman why her organization decided to recruit candidates under forty. She replied, "They're wildly underrepresented, and they directly affect the tenor of conversation on everything from housing to climate change to equal pay to student loans to child care. Also, when you elect more young people, you create a bigger pool of folks who can rise to the top."[19]

The roots of gerontocracy can be found in election law. In the first one hundred years of the United States, only white male landowners over twenty-one had voting rights. A public debate about lowering the voting age began during World War II. Proponents argued that young people could bring to politics idealism and an openness to new ideas. Opponents, by contrast, have argued that young people lack experience and good judgment. The fight for young people's democratic rights continues in the 2020s. Citing social research on the topic, two congresswomen stated in 2022 that "young people still participate in elections at lower rates than older age groups" and "routinely face serious obstacles to voter registration and in-person voting." Additionally, young people's "provisional ballots and mail-in ballots are rejected at disproportionate rates."[20]

Scholars of age bias in politics argue that there is a "vicious circle of youth alienation." A meager representation of youth in the political process results in decisions that disfavor youth policy preferences. This in turn leads to growing alienation from political participation and the idea in political parties that youth are less likely to vote and have influence. Gerontocracy further obstructs the process of political socialization among young adults who are motivated to participate and grow as politically reflexive citizens but are discouraged by the culture.[21]

Exclusionary Expert Culture

Another democratic problem is experts using their knowledge to exclude newcomers from the political field. Modern political institutions are complex and deal with complex problems, so newcomers can easily feel unqualified to participate. However, as demonstrated by Trump, politicians are not necessarily elected because of their level of understanding. Pierre Bourdieu has argued that professional politics has some of the same exclusionary mechanisms as other organizational fields. He writes that laypeople are led to believe that they intrude in a sacred space of politicians and lack legitimacy.[22] Laypeople therefore worry unnecessarily that they do not know enough or that they are in the wrong place, feeling that politics belongs to politicians. After all, how can ordinary people know where to find trustworthy political analysis and detect fake news? How can they know if the Biden administration's Inflation Reduction Act is a step forward in climate politics? New York socialists have spent years researching and deliberating complex questions about climate and democracy. They include, How can we combine movement and electoral politics? Should we run candidates on the Democratic Party's ballot? How can we finance political campaigns as a 501(c)(4) organization? What does the energy sector in New York State look like, and what is our plan for making it sustainable?

Elite Domination

Entering the political field is harder if you are poor, are a person of color, or do not have a college degree. Even if you are in a grassroots organization, you will be in a field where people in positions of power are overwhelmingly white and affluent and have a higher academic degree. The term "political class" refers to people with a level of power that makes it meaningful and possible for them to seek political influence. Lower-income, Black communities in New York generally feel ignored by political institutions and therefore have little faith in them.

The class structure of the political field has been the subject of much criticism among leftist thinkers. In the 1930s, the economist Joseph Schumpeter described democracy as a process of selecting leaders from among the elites.[23] A variation of this view can be found in Noam Chomsky, who describes US democracy as "a system of elite decision and public ratification." The influence of economic elites has grown with the deregulation of campaign finance law that happened in 2010.[24] A major aspect of elite domination is media ownership. Super-rich right-wing figures such as the Koch brothers, Rupert Murdoch, and Elon Musk own major news outlets and social media platforms. Habermas argued in the 1980s that commercial interests in news media led to distortions of democracy, and Nancy Fraser has argued that social inequality "taints deliberation within publics in late capitalist societies."[25]

The democratic socialist revival has emerged to oppose the vicious cycles in political life addressed in this chapter. The following chapters explore how.

2

The Sociology of Movement Parties

"Who gave you orders? I'll go after him. He's the one to kill."

"You're wrong. He got his orders from the bank. The bank told him, 'Clear those people out or it's your job.'"

"Well, there's a president of the bank. There's a board of directors. I'll fill up the magazine of the rifle and go into the bank."

The driver said, "Fellow was telling me the bank gets orders from the East. The orders were 'Make the land show profit or we'll close you up.'"

"But where does it stop? Who can we shoot? I don't aim to starve to death before I kill the man that's starving me."

—John Steinbeck, *The Grapes of Wrath* (1939)

John Steinbeck had a remarkable talent for capturing the working-class experience of alienation from capitalism. In *The Grapes of Wrath*, he wrote about a community of tenant farmers being displaced from their homes and the devastating consequences that followed, the looks of starving women and children as the men are informed about the eviction. Some of the men wanted to fight for their land, but they could not do what their families had done in the past—settling the matter through force in their immediate local environment (their ancestors had killed Indigenous people, among other actions). Industrial landowners posed a new kind of challenge. The workers did not understand how the owners could be so driven by a seemingly meaningless profit increase and destroy the community under the rule of law. They also did not understand how to take on the complex international institution of capitalism that inflicted such pain and suffering in their lives. These challenges are familiar to the millions of people who became poor during the Great Recession in 2008 but also to the growing number of victims of climate change.

This chapter reviews sociological theory on movement opposition to neoliberal capitalism in the wake of the Great Recession. It focuses on the international insurgency mobilized by democratic socialist movement parties. My argument in this book is that NYC-DSA can be situated in this history.

The chapter begins by introducing the subfield of political sociology, which helps explain how nonelites politicize their grievances and organize to counter elite dominance in political institutions. This perspective is essential to understanding the socialist and climate movements. The introduction to political sociology is followed by a review of the literature on socialist movement parties, with a focus on the relationship between ideology and organization in this insurgency. Next, I expand on this interpretation by revisiting key arguments in social movement theory and identifying the sources of conflict.

What Is Political Sociology, and Why Does It Matter?

In the most general terms, political sociology understands political life as the product of broader social and economic relations. The field has overlapping interests with political science, but the latter focuses more on the inner workings of national political systems, including aspects of electoral politics and government, such as voter behavior, public administration, and lobbying. Political sociology looks at politics more broadly as a place for settling conflicts in society and how political processes are shaped by social groups and capital. For instance, sociologists examine how political participation is structured by race, class, and gender, attending to the exclusionary mechanism of the political field and non-institutional alternatives. Moreover, sociologists examine how processes of politicization evolve across the spheres of everyday life and institutions. This is relevant to understanding why concerned climate citizens have primarily organized in social movements, not political parties, and why powerful politicians accept donations from the fossil fuel industry.

I draw on the so-called institutional tradition of political sociology that focuses on institutional politics and expand the perspective to include organized political life more broadly, including social movement organizations. The institutional tradition originated in the subfield's formative era of the 1950s. Lewis Coser distilled it into his 1966 definition of political sociology as "the sociological study of the social and political conflicts that lead to changes in the allocation of power. All study of political processes focuses attention on the state."[1] This formulation is indebted to Max Weber's conception of politics as the attempt to "influence the distribution of power, either among states or among groups within a state." For Weber, politics is an institutional process of resolving power struggles in society through negotiations that have a formal and legal status. He emphasized the authoritative status of the state. The state is "a human community that claims the monopoly of the legitimate use of physical force within a given territory," writes Weber. The strict focus on the government itself eventually became the field of political science, while sociology retains a focus on the social bases of politics in groups, organizations, and informal institutions. Even so, there is a history from Marx to Bourdieu for critical sociological thinking about the state in society.[2]

The emergence of political sociology as a subfield in the 1950s was partly a response to new social movements. It was also influenced by the deeper movement away from sociology's view of society as a relatively unitary entity marked by consensus toward a view of society defined by conflicts, including class tensions. Social movement research became a central area of political sociology, and movement scholars sought to expand and challenge the dominant norms in sociology that took the reproduction of social order for granted and understood politics in narrow institutional terms. Societal transformation became a normative ideal. Movement aspirations for change inspired critiques of dominant thinking about politics, identities, and institutions. Sociologists who were sympathetic to or involved with movements also reacted against the view of movements and protest as symptoms of social disintegration or societal pathology. They argued that movements are a normal part of politics, parallel to and interacting with parties and elections. Protesters are not inherently irrational and deviant. The new normative ideal can be registered in Elisabeth Clemens's 2017 definition of political sociology as "the sociological study of the emergence, reproduction, and transformation of different forms of political ordering."[3]

Political sociology has evolved since the 1960s with new theoretical perspectives and empirical orientations in response to new movements and societal changes. The Marxist wave of the 1970s inspired research on class conflict and social change, political revolutions, and working-class participation in politics. In the 1980s, Marxism gave way to poststructuralism and neo-Weberian thinking. Poststructuralism centered on Michel Foucault's theory of power and inspired critical thinking about gender and race. The exodus from Marxism led to the cultural turn, characterized by interest in meanings, morals, and emotions and to research on the middle classes. This development was inspired by Pierre Bourdieu's new theoretical perspective on inequality. Bourdieu understood class not merely as an underlying social structure but as a competitive process of striving. In focusing on how the privileged develop and retain their advantages, he demonstrated that the political classes—those with high economic and cultural capital—compete internally while sharing a dominant position over the popular classes. These advantages allow them greater opportunity to sustain their privileges. In the cultural turn, scholars began exploring how class is experienced affectively. In a study of working-class women, for instance, Beverly Skeggs found an emotional politics of class fueled by insecurity, doubt, resentment, and a sense of lacking alternatives.[4]

Sociological Responses to Late Neoliberal Capitalism

Neoliberalism is a broad and diverse phenomenon with diverse origins. The political scientist Wendy Brown argues that neoliberalism is commonly understood as a set of economic policies, including deregulation of capital, privatization of public goods, and the end of wealth distribution. She argues for a more complex understanding that includes Michael Foucault's idea of neoliberalism as "a normative order of reason," a political rationality that frames all forms of conduct as economic conduct in a profit-driven regime."[5] While I acknowledge the relevance of these analytical perspectives, they lend themselves to a somewhat abstract understanding of political history. As stated in the introductory chapter, I think that David Harvey has an important point in interpreting neoliberalism as a political project of economic elites. The neoliberal project emerged in concrete legislation in response to the financial crisis in the 1970s. The elites had leverage because the crisis had made political institutions economically and morally vulnerable. Harvey's interpretation does not prevent us from recognizing neoliberalism as a compelling response to growing individualism. Neoliberal policies were presented as a solution to a severe economic recession in the 1970s, and this pattern has continued since in what has become known as the "Long Downturn." Following an era of stable growth in both gross domestic product (GDP) and worker pay in the mid-twentieth century, the Long Downturn that began in the 1970s is characterized by severe ongoing recessions. Governments made emergency cutbacks in public spending and supported attacks on labor movements and reductions in real levels of pay.

Brown outlines four critiques against neoliberalism in the literature, including intensified inequality, unethical commercialization, growing intimacy of corporate and finance capital with the state, and economic instability. The political sociology of democratic socialist movement parties focuses on rising economic inequality and its implications for organized political life. How have decades of welfare cuts and severe recessions shaped ordinary people's relationship to political institutions? What are the implications of rising oligarchy?

The historically high level of economic inequality has led to a broader wave of interest in this issue across the social sciences. In sociology, there is a revival of interest in basic material issues and in Marxism. Issues of class and capitalism have returned to the center of public debate and critical sociological research. The new leftist movements after 2008 have helped transform political perspectives in academia and the general public, which is now much more sensitized to anticapitalist narratives.

The revival of interest in basic material issues is a return to one of the foundational motivations of sociology in the nineteenth century: the political critique of neoclassical economics. University programs in capitalism studies are emerging, and there is a new wave of Marxist scholarship pioneered by scholars such as Thomas Piketty, Nancy Fraser, Richard Wolff, David Graeber, Vivek Chibber, and Matt Huber. This literature updates Marxism for our time and takes issue with liberal conversations about inequality. Graeber and his coauthor David Wengrow point out that inequality became recognized as a root problem in society after the Great Recession, even among the liberal political class.

> Today, there is a veritable boom of thinking about inequality: since 2011, "global inequality" has regularly featured as a top item for debate in the

> World Economic Forum at Davos. There are inequality indexes, institutes for the study of inequality, and a relentless stream of publications trying to project the current obsession with property distribution back into the Stone Age. . . . The ultimate effect of all these stories about an original state of innocence and equality, like the use of the term "inequality" itself, is to make wistful pessimism about the human condition seem like common sense.[6]

Graeber and Wengrow suggest shifting the conversation from inequality to the concentration of capital and class power. They also believe that there are better alternatives to the current economic system and therefore encourage experimentation with alternative forms of social organization. Such experimentation has always been part of human history, but it is lacking today because neoliberalism suppresses alternatives to capitalism.

Tom Piketty has drawn attention to the historical transformation in political consciousness and highlighted key aspects of contemporary Marxism, including the focus on feminism and ecology. He argues for a transformation of the economic system that takes other crises into account and draws lessons from social movements.

> If someone had told me in 1990 that I would publish a collection of articles in 2020 entitled *Vivement le socialisme!* in French, I would have thought it was a bad joke. As an 18-year-old, I had just spent the autumn of 1989 listening to the collapse of the communist dictatorships and "real socialism" in Eastern Europe on the radio. . . . But now, thirty years later, hypercapitalism has gone much too far, and I am convinced that we need to think about a new way of going beyond capitalism, a new form of socialism, participative and decentralized, federal and democratic, ecological, multiracial, and feminist.[7]

The Marxist revival has not yet fully arrived in the social movement literature in the United States. This literature is still in the process of refining older paradigms and catching up empirically with new movements, particularly the climate movement and the Christian nationalist movement. Scholars such as Michele Lamont, Dana Fisher, and Theda Skocpol implicitly adopt a liberal-progressive perspective. Lamont explores narratives of justice in civic society without analyzing their political economy. Fisher has written about the 2017–2018 anti-Trump movement, which was primarily populated by people who had voted for Hillary Clinton, without considering the democratic socialist movement led by her opponent Bernie Sanders, who is wholly focused on economic inequality. Fisher, Skocpol, and others provide valuable insights into the anti-Trump movement that remain relevant for situating the democratic socialist movement. The anti-Trump literature shows that many of the progressives who volunteered for campaigns organized by the Democratic Party were not close to the party. Many of them were disappointed with the party, but they were motivated by negative partisanship, the eagerness to support an alternative to Republican Party dominance. These self-organized progressive grassroots in small cities and suburbs did not create their own organization. This partly explains why these groups died out so quickly. By contrast, NYC-DSA organized its own elections. This helped the socialist anti-Trump movement in New York City grow its institutional base. NYC-DSA won its first campaign for state government weeks after the 2018 midterms, and it continued to win more elections in 2020 and 2022. This development has not yet been subject to scholarly analysis.[8]

Next, we turn to research on socialist movement parties in Europe and Latin America to provide a theoretical and historical framework for understanding NYC-DSA since 2016. This international perspective helps overcome the typical national framing of movements and the limited attention to economic inequality in US movement scholarship.

Socialist Movement Parties

Protests in the Wake of the Great Recession

The international wave of protests against austerity policies and economic inequality has inspired new thinking about the relationship between movements and parties in political sociology. The protests began in Latin America in the 1990s and developed in Europe a decade later in response to the Great Recession.

The Italian scholar Donatella della Porta and colleagues developed comparative international perspectives on southern Europe, where the Recession hit hard. These scholars argue that the Recession was a critical juncture that triggered socioeconomic and political transformations. At the center are protests against the austerity policies adopted by national governments. For this reason, they use the label "anti-austerity" movements. I call them "socialist" to identify their political ideology. The protest claims were largely about restoring classic socialist principles. They opposed the retrenchment of the welfare state under neoliberalism and the corruption of democracy by the 1 percent. They further sought to unite mass numbers of citizens experiencing a drastic decline in living standards.[9] These ideas are central to the democratic socialist movement in the United States, including its arguments for universal health care, labor rights, and a Green New Deal.

In southern Europe, the protests evolved into parties shaped by the particular situation in each country. In Greece, the SYRIZA party was connected to the European communist Left and responded to the issue of national debt. SYRIZA emerged as a coalition of parties in 2004 and reregistered as a single party in 2012. It went from getting 4.6 percent of the vote in 2009 to 36.3 percent in 2015. Following this victory, SYRIZA became the ruling party and the only left-wing party to win government in Europe. In Italy, the Five Star Movement party (Movimento 5 Stelle; M5S), created in 2009, was defined by the charismatic leadership of the comedian and activist Beppe Grillo and focused on the country's declining productivity. Italy's GDP growth had been among the lowest in the Organisation for Economic Co-operation and Development (OECD) for decades. In the 2012 local elections, M5S got more than 10 percent of the vote in many northern areas. It became the second-largest party in the national election in 2013. In Spain, the Podemus party was defined as an alternative for the social democratic electorate disaffected with the Socialist Party and mainstream political parties more generally. It emerged in 2014 from anti-austerity mobilizations, beginning with the Indignados movement in the early 2010s. Podemos was launched by a group of activist-scholars with experience in the global justice movement and people involved with the Spanish radical Left. Its popularity peaked in the 2015 national election, when it became the third-largest party.

All three parties mobilized support around the economic crisis and the lack of trust in political parties. They emerged from a space left by the dramatic decline in support of the traditional center-Left, which was no longer perceived as an alternative to the center-Right. Still, how could protests evolve into big parties in the time span of a decade? There was a dramatic decline in household incomes and support for established parties. In Greece, for instance, household incomes decreased four times the average of Eurozone countries, and two major parties collapsed. Austerity measures led to cuts in salaries and pensions, to the extent that many people with higher education became homeless. Most supporters of the new parties were in their twenties and thirties, well educated, and from big and medium-size cities. The demographic generally became broader and more heterogeneous as the movements evolved into parties. Podemos had a younger and more urban profile than SYRIZA did. Podemos was the most popular party among young people, with 27 percent support in the eighteen to twenty-four age group and 20.6 percent support in the twenty-five to thirty-four age group. SYRIZA and Podemos have declined considerably in the early 2020s, and only a few European governments are currently led by a left-wing party. The Right gained further dominance in the elections for the European Parliament in 2024, but this rightward shift creates opportunities for the Left. The second Trump administration has stimulated European skepticism about the far Right and oligarchy.[10]

The Decline of Political Party Organizations

The rise of movement parties happened in the context of a longer transformation of political parties in neoliberal capitalism. The transformation was political and organizational. Herbert Kitschelt reminds us that the demise of universalist social democracy in the late twentieth century was related to the changing class structure of advanced capitalist societies and their development into welfare states. In Europe, the development of comprehensive welfare states dismantled the programmatic alternatives for distributive politics, and this safety net paradoxically contributed to a new form of individualism. The welfare state made the citizen less dependent on the family but also institutionalized individualized responsibility in the context of capitalism, contributing to the normative ideal of living a life of one's own.[11]

Kitschelt further links growing educational levels to transformations in political life. New classes of educated voters mistrust parties and are ready to defect and switch to competitors if their concerns are not serviced. Parties move from being encompassing "department stores" to becoming "boutiques." Another change in parties is their increasing reliance on public financing to become independent of membership contributions. Parties are decoupled from their social base and rely more exclusively on their institutional base and professional public relations. This explains why parties converge on policy issues, even when voters do not converge on the same issues.[12] These are the core characteristics of the so-called cartel party model, which della Porta and colleagues take to be typical of parties in the neoliberal era. The model is further characterized by the centralization of power in the hands of a few leaders, with the mere formal involvement of rank-and-file members. This involves a growing focus on individual leader personalities over ideological framings. The relationship between parties and civil society has weakened with regard to both numbers and functions of party members.

> In sum, the emerging model of party presents a shallow, weak, and opportunistic organization; ideological appeals are (at best) vague, with an overwhelmingly electoral orientation. Electoralist parties debouch into personalistic parties, whose "only rationale is to provide a vehicle for the leader to win an election and exercise power . . . an organization constructed or converted by an incumbent or aspiring national leader exclusively to advance his or her national political ambitions. Its electoral appeal is not based on any programme or ideology, but rather on the personal charisma of the leader/candidate, who is portrayed as indispensable to the resolution of the country's problems or crisis."[13]

The socialist movement parties of the Great Recession have reacted against this organizational culture of political parties. They view social movements as a resource for rebuilding political trust and overcoming the growing gap between political institutions and voters.

The Great Recession also gave rise to populist right-wing movements. These movements compete with leftist movements in proposing solutions to downward mobility and the alienation from mainstream parties. The populist right-wing movements blame ethnic-racial minorities and immigrants for social problems and argue that these groups are draining society of resources they do not deserve. (The latter is known as ethnonationalist welfare chauvinism.) The primary social base of these movements is in rural and working-class communities deprived by globalization, but they are also supported by economic elites. They are led by demagogues who exploit the lack of trust in institutions and further weaken institutions.[14]

The Concept of a Movement Party

The socialist movement parties in southern Europe have inspired scholars to think about the potential of movement parties for overcoming the crisis of democracy. This perspective is explored by della Porta in her 2020 book *How Social Movements Can Save Democracy: Democratic Innovations from Below*. The core argument in much of della Porta's work throughout the 2010s is that movement parties are relevant to restoring popular political participation and improving the capacity of parties to mediate interests in society. In a sense, the new movement parties are filling a gap left by old parties, which have become estranged from large parts of the constituency.

Della Porta challenges conventional thinking about movements and parties as separate and mutually exclusive entities. She acknowledges that movements have historically ceased to exist as movements once they developed into political parties. The labor movement that gave rise to social democratic parties in the nineteenth century, for instance, has slowly eroded. However, in the current crisis of democracy, movements have gained renewed relevance to political parties. The hybrid movement party organizations of the Great Recession might have more long-term relevance and not necessarily be a transitional phenomenon like the green parties in the late twentieth century (see the introduction). One of the lessons from research on movement parties over the past decade is that movement theory has focused too narrowly on movements in themselves and neglected their relationship with parties.

In Kitschelt's formative theory of the 1980s, he describes movement parties as coalitions of political activists who are coming from movements and applying the movement organizing principles in the arena of party competition. Movement parties integrate movement constituencies in their organization. Kitschelt further argues that movement parties are likely to appear where (1) a large constituency is willing to articulate their demands through disruptive, extrainstitutional activities, (2) existing parties make little effort to embrace these interests, and (3) the formal and informal thresholds of political representation are low.[15]

Della Porta and colleagues argue that new socialist movement parties support a participatory vision, have a more decentralized organization, and promote new subjectivities. Additionally, several of their electeds have gained fast access to government and do not come from social backgrounds typical of politicians. Crucially, the new movement parties have emerged in the context of the accelerated crises of neoliberal capitalism, democracy, and climate change.[16]

A Social Movement Theory Perspective on NYC-DSA

Social movement theory can help explain the movement dimension of NYC-DSA and situate this organization in the democratic socialist movement. NYC-DSA is a grassroots movement-based organization. It has nine electeds in state government, but that is not its primary organizational base. NYC-DSA is run by the grassroots in the city, organized into neighborhood branches and citywide leadership committees (see chapter 7).

NYC-DSA is also as a movement organization because it is a platform in the democratic socialist and climate movements. A social movement is characterized by the social dynamic of a struggle shaped by a collective identity that exceeds the boundaries of any single group or organization. Mario Diani explains: "[A social movement is] a process whereby several different actors, be they individuals, informal groups and/or organisations, come to elaborate, through either joint action and/or communication, a shared definition of themselves as being part of the same side in a social conflict."[17]

The dominant framework in social movement scholarship of the past fifty years is the so-called resource mobilization theory (RMT). The framework informs my analysis of mobilization and collective action in NYC-DSA in subsequent chapters. It was defined by the work of John McCarthy, Mayer Zald, Charles Tilly, and Doug McAdam in the late 1970s and early 1980s.[18] Some of the foundational ideas were articulated by Charles Tilly in his book *From Mobilization to Revolution* of 1978.

Tilly observed the fragmented and underdeveloped state of thinking about collective action, protest, and movements and aimed to create the groundwork for a clearer understanding. He proposed a dual framework of mobilization and opportunity. The idea is that collective action is made possible by (1) the mobilization of resources (money, volunteers, leadership, etc.) and (2) the perception of opportunities resulting from changes in the national political environment, such as the weakening of political parties or government. The role of threats has been added later, and this is relevant to climate change.

The RMT framework challenged three dominant views of collective action at the time. First, RMT challenged the functionalist view that collective action is an outburst of uncontrolled behavior. It argues that individuals are purposefully participating with some level of rational consideration of their interests. Second, RMT challenged the deprivation theory of collective action, which sees it as being a result of failed expectations. And third, RMT challenged the material view that collective action emerges from structural tensions, such as growing class divisions.

One inspiration for RMT was Mancur Olson's *The Logic of Collective Action* (1965), which took a cost-benefit approach. Olson considered why people decide not to participate in collective action. He argued that many people think that their individual participation makes a small difference to an action's success relative to the personal costs they would incur. Also, they know they can benefit from group gains either way. This theory was critiqued during the cultural turn in the 1980s for being overly rationalist and structuralist. People are not atoms but make choices with others and gain emotional and social benefits from participating in collective action.[19] This view resonates with my experience in NYC-DSA.

In the cultural turn, David Snow and collaborators developed a theory of narrative mobilization, drawing from Erving Goffman's theory of framing. The argument is that movement organizations create narratives that motivate people to participate by making them see the world in a new way, using symbols and stories that highlight the urgency and morality of a social problem. The narrative frames are devices with which organizations create shared understandings and interpret events. They help participants understand themselves as part of the movement and thus create collective identities and cohesion in movements. This perspective has since been expanded to include other modalities of communication: Movement actors are using visual means such as movies, murals, effigies, and costumes as carriers of meaning.[20]

In the affective turn, James Jasper has argued that emotions are essential to how movements conduct their central task of public persuasion and move people into action. Movements depend on feelings of trust, admiration, and loyalty to the collective. Moral shocks can push people into action. Moral emotions such as pride, indignation, and anger direct attention to social problems.[21]

Narrative and emotional mobilization are evident throughout NYC-DSA campaigns. The campaigns define climate change as a crisis and formulate a positive alternative, a socially just and green New York. They create community among volunteers through movement leadership and use the symbolic colors of socialist red and climate green in campaign visuals to make the identity and experience more powerful (see chapters 4 and 9).

Finally, social movement theory has addressed the historical dimension of how movements adapt to social change. Tilly's history of contentious collective action in France since the late seventeenth century shows how organizations and tactics have evolved in response to developments of capitalism and the nation-state. As the national government became more powerful, protesters directed their claims toward elected officials instead of burning down a local mansion or tar and feathering a tax collector. In the 1700s, protesters began targeting the imposition of new forms of state and capitalist control, using a parochial repertoire grounded in local routines and folklore and relying on patronage. As power became more concentrated in national government, large organizations, and big cities in the nineteenth century, protest became structured by specialist national organizations, and the repertoire evolved into national strikes, demonstrations, and electoral rallies.[22] I draw inspiration from this perspective to interpret NYC-DSA as a particular development from the anti-Trump movement and the socialist insurgency in neoliberalism but also the longer historical transformation of New York City political life in the twentieth and early twenty-first centuries. It would be simplistic to interpret NYC-DSA as a tribal response to a systemic crisis in modernity. Also, I emphasize the relevance of the organization's attempt to rebuild participatory democracy from the ground up after decades of neglect.

The Main Source of Conflict: The Crisis of Inequality

Tilly's thinking motivates attention to the historical sociology of movement parties. To this end, I return to della Porta. She argues in *Social Movements in Times of Austerity* (2015) that the socialist movement parties of the Great Recession are not merely protesting the consequences of neoliberal capitalism but are shaped by them: The movements emerge from the new precariat created by neoliberal capitalism, and the conditions and demographics of this precariat shape the outlook of the socialist movements and their relationship with the state. Neoliberal capitalism has produced a precariat of young, educated, but unemployed and underemployed citizens with minimal trust relations with capitalism and the state. This precariat does not have the privileges of the managerial class or the securities of the historical working class. These structural changes explain new alliances between educated young urbanites and wider classes and age groups in society.[23] Della Porta further suggests that the socialist movements are shaped by a crisis of legitimacy and responsibility, which I extend to include government climate neglect (see chapter 1). Neoliberal capitalism involves an immoral dimension, with a cynical refusal of values of social protection and solidarity. Governments have directly undermined labor unions, and unions have adopted corporate management styles that weaken their connection with the everyday experience of workers. Movements have responded to these issues with appeals to reestablish the social order that they perceive as broken. The transformation of unions further explains why they have not provided a platform for the socialist movement.[24]

One of della Porta's key arguments is that the new socialist movements evolve from growing social cleavages. In general terminology, "political cleavages" are enduring group differences, typically defined by class and ethnicity, that are mapped onto party alternatives. "Social cleavages" are group divides in social life as a whole, and they do not necessarily translate into political cleavages. The socialist movement evolves from growing structural economic inequality and seeks to map this issue as a party alternative. The new democratic socialist parties use the class concept to frame their struggle and identity. Thus, there is little doubt that the massive changes in wealth distribution over the past four decades are relevant to the analysis of movement narratives and organizational practices in political life. This motivates revision of the RMT framework, which evolved when Marxism was losing ground in the 1970s. The framework focused on the process of mobilization, not the formation of political positions and their underlying structures. Chapter 3 provides insight into new narratives in the current political environment.[25]

3

The Political Environment After the Great Recession

Like all movement parties, the new NYC-DSA of post-2016 arose in response to developments in the national political environment. This chapter investigates how the people involved in the organization's transformation responded to key national events from 2008 through 2016. It draws from the resource mobilization framework in social movement theory (see chapter 2) to explain how my informants interpreted these events as crises, opportunities, and solutions.

I quote extensively to capture this formative moment in the organization's history from those who experienced it. The chapter highlights the perspectives shared by my informants, providing ground for the exploration of diverse perspectives among participants in chapters 6 and 7.

Media shaped the movement's narratives, but I analyze conversations held with informants, not media content. My focus is on how participants experienced the political environment. Another reason for this approach is that several informants explained that they had deliberately moved away from reading political journalism to concentrate on collective action. They were generally not avid readers of political journalism, nor were they interested in talking about it. They did not regularly browse major news websites, and they felt animosity toward these news outlets. A future media reception analysis could add nuance to this exploration.[1]

Opportunities, Crises, Solutions

One of the core ideas in the RMT framework (see chapter 2) is that movement actors view changes in the political environment as opportunities for mobilization. The actors interpret negative trends, such as growing inequality and climate change, as symptoms of crises. The crisis narrative can help recruit people experiencing social frustration and move supporters into collective action. It can shape people's perspectives on their grievances and create attachment to the collective identity of the movement. In the textbook version of this process, the movement narrative is promoted by movement organizations—in this case NYC-DSA. In the US democratic socialist movement, the influential narrative has come from political opinion leaders, especially Bernie Sanders and leftist independent media. DSA has not asserted itself as a voice in the national political public sphere. The "new DSA" emerged when the general narrative of democratic socialism was already in place, and it does not focus on national politics. The perception is that the organization is too small to contest for power at the national level and that national politics is at an impasse. DSA is a decentralized organization with each chapter organizing with a local focus. NYC-DSA's external communications primarily take the form of issue campaigns and electoral campaigns targeting the New York State government.

The rise of democratic socialism can be explained as the culmination of a decade of evolving crisis-and-opportunity narratives: I identify the Great Recession as a landslide event for cascading leftist opposition to neoliberal capitalism. The Great Recession, of course, had long-term social and political implications, provoking fundamental issues that would accelerate over the next decade: growing income inequality, declining labor conditions (increasingly influenced by technology corporations in California), and the election of a fascist president in 2016 who has since continued to wreak havoc.

As for the opportunities, early events include Barack Obama's 2008 presidential campaign and the Occupy Wall Street protests in 2011. A deeper transformation evolved from the 2016 Sanders campaign. This campaign tapped into the radical spirit of Occupy and revived the use of the word "socialism" during the Democratic primary (see introduction).

TABLE 3.1. National Events That Shaped the New NYC-DSA

Year	Event	Opportunity	Crisis
2008	The Great Recession		Economic crisis and inequality
	Obama's campaign and victory	A progressive, movement-oriented candidate with a message of hope	Obama's abandonment of progressive campaign goals
2011	Occupy Wall Street	An anticapitalist insurgency against neoliberalism	
2016	Sanders's campaign	A socialist insurgency in the Democratic Party	Sanders's loss, facilitated by the national Democratic leadership
2016	Trump's election	The Democratic Party's weak response to Trump	The moral shock from the election of an autocrat backed by a far-right movement
2018	The Blue Wave and the Squad	National campaigns such as Tax the Rich and the Green New Deal	

TABLE 3.2. Socialist Members of Congress

	2019	2021	2023	2025	State or district
Bernie Sanders 1991–present	X	X	X	X	Vermont
Alexandria Ocasio-Cortez*	X	X	X	X	New York City
Ilhan Omar	X	X	X	X	Minneapolis
Ayanna Pressley	X	X	X	X	Boston
Rashida Tlaib*	X	X	X	X	Detroit
Jamaal Bowman*		X	X	—	New York City
Cori Bush*		X	X	—	St. Louis

* = DSA member for a few years or more

TABLE 3.3. DSA-Backed Electeds in New York State Government

	2019	2021	2023	2025	Area
Julia Salazar	X	X	X	X	Brooklyn
Jabari Brisport		X	X	X	Brooklyn
Marcela Mitaynes		X	X	X	Brooklyn
Phara Souffrant Forrest		X	X	X	Brooklyn
Emily Gallagher		X	X	X	Brooklyn
Zohran Mamdani		X	X	X	Queens
Kristen Gonzalez			X	X	Queens
Sarahana Shrestha			X	X	Hudson Valley
Claire Valdez				X	Queens

Crisis 1: The Great Recession (2008)

My informants generally did not address the Great Recession, but I include it here because of its importance to the events that followed. The Great Recession was a peak moment in the so-called Long Downturn and the worst recession since 1929 (see introduction and chapter 2). It exposed the extreme inequalities of neoliberal capitalism: Wall Street received multibillion-dollar bailouts and distributed $18.2 billion in bonuses, while 2.6 million jobs were lost and 6 million people lost their homes to foreclosure. The US middle class shrank to around 50 percent, down from 60 percent in 1971. It was squeezed with regard to gas prices, medical insurance, and college costs. At the same time, more than half of all corporations were not paying taxes.[2]

Opportunity 1: The Barack Obama Campaign in 2008

The 2008 Obama campaign created a glimmer of hope for progressive politics and youth grassroots participation. It effectively deployed a movement organizing approach, which inspired a new trend in civil society organizing and eventually shaped NYC-DSA.

One of NYC-DSA's most influential organizers, Tascha Van Auken, worked for the Obama campaign. This was a formative experience:

> I was very interested in the movement and energy that was developing around the Obama campaign. And so I wanted to volunteer but basically got lucky and got hired instead to work in a field office in Pennsylvania for the last two and a half months. I loved it. His 2008 campaign was really well run. And they dumped a ton of money and spent money on smart things. So they had a million field offices, hired a ton of organizers, trained people really well, and set the bar high for me in terms of what a good field operation is. It was all about elevating volunteers into leadership roles. When I got there, everything was already being run by middle-aged women across the district. They were doing all the trainings. They were the ones who knew everything, and so my job was just to, like, get the office into shape and get it organized, so that all of these people coming in had work to do. The methodicalness of it was really appealing to me, the structure, and the goal being

to just talk to voters. This was the first time I felt I was part of a community and got pulled into this community just really fast. I still talk to these people.³

Yet the excitement quickly dissipated once Obama took office. Many young leftists were disappointed by his adoption of neoliberal values and hawkish foreign policy. The disappointment was stronger because Obama had acted like a visionary. He represented the growing centrality of public relations in politics and its role in the crisis of democracy. Since the 1960s, political leaders in the US have become increasingly beholden to campaign specialists and less so to their party and voters.⁴

> JULIA SALAZAR: I went to college [at Columbia] in 2009 when Obama had just been elected in his first term, and that was really exciting. It did create this environment that was empowering for progressive-minded people. At the same time, we were at war. So that was disillusioning.
>
> DAVID ALEXIS: Obama's election was the first election I really paid attention to. I was just old enough to vote for him. I was excited about Obama, but I was very disappointed with what happened after.
>
> SAM LEWIS: The disappointment of Obama's first term was enormous.
>
> WILLIAM RUDEBUSCH: You listen to Obama and, later, Hillary and Biden, and they're just like, "America is about opportunity." What do these words mean? These are just happy-sounding words. They don't offer answers. I was listening to this podcast where they asked Democrats about their opinion on Obama. If someone said anything even slightly negative . . . I mean it's completely ineffectual. But oh, "Obama was just such a good speaker." He was still bombing Syria!⁵

Opportunity 2: Occupy Wall Street in 2011

Occupy primarily inspired the democratic socialist movement by signaling a general anticapitalist perspective on extreme inequality. The statements about the "1 percent" and "people over profit" are now commonplace in social movements around the world.

Michael Paulson, a climate organizer in NYC-DSA since 2017, remembers Occupy as an important politicizing experience:

> PAULSON: I did not grow up in a very political household at all, and I was not super engaged until my late twenties. I was doing a PhD in English, and during that time, I got exposed to more political writing and people who are more political. The experience that really got me involved was the unionization drive at Columbia University. I was one of the first generation of organizers of that union, and that basically made me into an organizer. It also just made me much more aware of politics as a contestation of power—and not only in the electoral arena.
>
> AUTHOR: What are some of the national political events that have shaped your political identity?
>
> PAULSON: There were major events that didn't politicize me, like 9/11 and the Great Recession. But Occupy . . . It's easy to critique Occupy as, like, "Okay, this is not connected to building power," but on some level, it was just an act of pure negation, which did cause people to just think, also about the concept of politics. I grew up in the '80s and '90s, where it was the neoliberal heyday. There was no political choice to be had. It was just, "You're a consumer." There wasn't a Left. And the parties were not really polarized in the way that they are now. Some people got very involved in antiwar activism or antiglobalism, but there was no unified political analysis and program in the way that a socialist party would have.[6]

Neal Meyer remembers how people in NYC-DSA experienced Occupy as a major opportunity. He had been a member of the youth affiliate of DSA (YDSA) along with Bashkar Sunkara, who edited the organization's magazine and started *Jacobin* in 2010. This magazine quickly became the leading publication in the emerging urban socialist movement, and Meyer and Sunkara developed and guided the hundreds of *Jacobin* reading groups all over the country and even internationally. These groups met once a month to discuss *Jacobin* articles and other socialist literature. Eventually, the reading groups created a social base for DSA. Most of them converted into DSA chapters. Meyer remembers,

In 2013, I was a YDSA organizer. At the time, it was really difficult to get people to join DSA or even express interest in joining DSA. At the beginning of every school year, the YDSA organizer would get maybe thirty to forty people reaching out from different colleges to say, "Oh, I'm interested in starting a DSA chapter." And we would obsess over those contacts! We would be really concerned with each one of them and trying hard to get them to start a DSA chapter. Some of them will do a great job. With others, it would be like pulling teeth to get them to do anything. And I became kind of frustrated with that situation. I started talking to Bhaskar about frustrations I had meeting a sufficient number of people to really get a project going, and we were talking about "How are we gonna get the socialist Left organized and back on its feet?" We're just kind of thinking about how many contacts the *Jacobin* had. At the time, the magazine already had thousands of readers and a really engaged readership that was active online and stuff. And he knew about reading groups of *The New Left Review* in the 1960s and was thinking about those. There were other precedents. I guess *The Nation* had reading groups. And Bhaskar had actually suggested the idea of doing reading groups years prior. We had thought about it, and it didn't go anywhere. But basically, we kind of had a more serious conversation about it in the winter of 2013–2014. And then he offered me a job. He was, like, you know, "Come work for us." I was the part-time reading group organizer, part-time circulation manager. So half the time I was just kind of filling mail orders and stuffing envelopes. The other half I was doing organizing.

Our idea was basically that there was this new opening post-Occupy. There was a small but critical number of millennials who were interested in socialist politics, and we wanted to help them deepen their understanding of what it meant to be a socialist and what socialism is all about. And that the reading groups would be this kind of really easy to access space that was not sectarian, not attached to any of the old socialist groups, including DSA. They would be open to anybody who had an interest in socialist politics, and we wanted them to be really easy to organize and to have some support.[7]

Several DSA organizers felt that Occupy was limited by poor organization. Van Auken explains,

I got involved with Occupy and mostly didn't have a good experience. I've mostly learned about what not to do by trying to get involved with Occupy [laughs]. It was very dysfunctional—very much elevated toxic personalities who wanted to control things. There wasn't structure. There wasn't accountability in mind. The ideas were great, and it was a really great moment because you could see, "Oh, there are thousands of people that want to get involved!" But when you went to the occupation and tried to get involved, it was very hard. You felt very much that it wasn't inclusive. I went to dozens of meetings and couldn't penetrate the structure.[8]

Marsha Niemeyer, a longtime union organizer who joined NYC-DSA in 2018, suggests that Occupy reflects a general decline in civil society organizing and had an element of organizational anarchism:

It was clear that there was already something bubbling up under the surface before Bernie. Occupy helped with that a little bit, but it also illustrates the deep deficit of collectively organized spaces in US society for a long time. I mean, the primary collectively organized space that continues to exist is religion. Maybe sports too. Certainly, organization was not what they were grappling with at Occupy. They were loosely trying to figure out how to organize themselves, but it wasn't put in place to have any lasting impact. It wasn't put in place to have any semipermanent structures come out of it or even semipermanent leaders come out of it. It was extremely disorganized. It was an amazing explosion. There were Solidarity [a revolutionary socialist organization] people here who were just tearing their hair out trying to figure out how to intervene and how to cohere. It was the wrong reaction to have: Occupy wasn't something you could intervene in.[9]

Occupy represents what some observers call "reactive organizing" because of the emphasis on protest and short-term organizing for immediate, direct action. They contrast it with the "proactive organizing" of developing a long-term political project with corresponding organizational structures. My informants talked about a historical shift away from the reactive culture of the alter-globalization movement of the 1990s to the constructive electoral project of the democratic socialist movement. Said the NYC-DSA organizer Josh Kraushaar, "Occupy was an older phenomenon than whatever we're trying to do now, especially post-2016. You need to build power. Electoralism is attractive because it's clear what power is in that context. It's limited, but at least it's concrete. Occupy did not really have cohesive demands. It was not focused on building power in any coherent way. It's this idea of fighting the system without any vision of what that fight really entails."[10]

Opportunity 3: The Bernie Sanders Campaign of 2016

The 2016 Sanders campaign transformed the political imagination of the Left by demonstrating the mass popularity of democratic socialism and the opportunity for socialist influence in the Democratic Party. The campaign motivated young leftists to enter electoral politics, thus stimulating the shift from reactive to constructive organizing.

Sanders was an experienced politician who stood up against the Democratic Party's alliance with the economic elites. He put economic inequality front and center of his presidential primary campaign, framing it as an alternative to the centrist party line that was consolidated during Bill Clinton's presidency in 1993–2001:

> Hillary Clinton was a key player in the centrist Democratic establishment, which had, over the years, been forged by her husband, Bill Clinton. In fact, Bill Clinton had been the head of the Democratic Leadership Council (DLC), a conservative Democratic organization funded by big-money interests, which was described by Jesse Jackson as "Democrats for the Leisure Class." The Clinton approach was to try to merge the interests of Wall Street with the needs of the American middle class—an impossible task. . . .
>
> For me, the bottom line was that this country was facing enormous crises: the continued decline of the middle class, a grotesque level of income and wealth inequality, high rates of real unemployment, a disastrous trade policy, an inadequate educational system, and a collapsing infrastructure. On top of all that, we needed bold action to combat climate change. . . . Did I believe that the same old establishment politics and establishment economics, as represented by Hillary Clinton, could effectively address these crises? No. I didn't.[11]

Leftist media interpreted Sander's campaign as a major opportunity structure. Many articles in *Jacobin* and *Dissent* asked the underlying questions, "Could he win?" and "What will happen to the Left if Sanders is no longer a presidential candidate?"

Sanders became *the* heroic figure for the democratic socialist movement. His biography lent credibility: He grew up in a working-class community in Brooklyn, he was an activist in the civil rights movement with the Young People's Socialist League, and he framed his campaign as an organic development from a long career as an independent politician with moral integrity, from city mayor to US senator. This was not a person who came up through the ranks of the Democratic Party. Many people in NYC-DSA have read his autobiography, *Outsider in the White House*. He has long positioned himself as a David fighting the Goliath that is corporate America and frequently mentions that he does not accept money from corporations—a principle that DSA adopted. One event that contributed to his identity as a heroic insurgent was his eight-and-a-half-hour speech on the Senate floor in 2010. In this speech, he opposed the tax agreement between Obama and the Republicans.[12] He continues to appear at union strikes and picket lines around the country and recently attracted huge crowds to his "Fighting Oligarchy" rallies with AOC. The appreciation for Sanders in NYC-DSA showed in many situations, one of which is particularly memorable: When a campaign manager in NYC-DSA faced great challenges in 2022, he posted on social media, "When I find myself in times of trouble, father Bernie comes to me," and he continued paraphrasing the entire verse and chorus of The Beatles song "Let It Be." There were many "Bernie Bros" in NYC-DSA in 2016–2018, but the chapter was significantly more gender-balanced than many other chapters and has since become more culturally diverse.

How did NYC-DSA members talk about the Sanders campaign as an opportunity?

> MEAGAN DAY: I didn't know I was a socialist until Bernie Sanders's first presidential campaign. I knew I was repulsed by exploitation and oppression, and I even understood that capitalism perpetuated much of the injustice I saw around me. But I had never even once considered the possibility that I myself was a socialist. No one had ever asked.

JENNIFER ROESCH: The struggles over the last two decades—especially the last eight years—are enormously important and paved the way for Sanders's success. It's almost a time-delayed "Occupy goes to the polls" kind of expression.

NEAL MEYER: Basically what happened is that at the end of 2015, myself and three other people from the *Jacobin* reading groups were like, "Okay, DSA is taking off. Bernie is creating a new set of conditions for democratic socialist politics." We're excited about that. We felt the DSA was the natural home for this kind of new generation. I had been a member of DSA since 2012, but very few were involved in the organization. Bernie gave me new hope that DSA could be a center of gravity for this new upsurge. We were like, "Okay, this is the place we want to help try to organize this kind of new generation into."

DEVON MCMANUS: I got activated by the Bernie 2016 campaign and joined DSA in 2017. I think Hillary losing to Trump and the sort of safe choice isn't working . . . I might as well go for what I'd actually like to see. I think the label "socialism" played a role because DSA is in the place it is today because Bernie Sanders promoted democratic socialism.

MICHAEL POLLAK: Bernie completely shocked us. I mean, sure, we supported him, but we didn't think he'd go anywhere. But he did, and he showed us that we could collect small donations via the internet and knock on doors. Then DSA completely transformed.

JOSH KRAUSHAAR: The shift that Bernie Sanders really brought around—and the emergence beyond it—is the very beginnings of a potentially viable Left politics and thinking about politics in terms of building power.

MICHAEL PAULSON: Sanders really made social democratic politics available to people in a way that simply was not thinkable before.

DAVID DUHALDE: Bernie Sanders made it much easier for people to run as socialists.

GUSTAVO GORDILLO: My guess has always just been that DSA grew in numbers in large part because Bernie Sanders talked about democratic socialism and a lot of people Googled it and DSA came up.[13]

John Tarleton, the activist and editor of *The Indypendent*, New York's only free left-wing newspaper, brings nuance to the picture based on decades of experience on the New York Left. Tarleton ascribes importance to Sanders's ability to offer a concrete democratic socialist interpretation of the living conditions for the working class, with compelling suggestions for a better life:

> Occupy definitely crystallized a certain kind of class consciousness, but the Sanders campaign, in particular—I mean, his sort of genius was and is, he emphasized these redistributionist issues, like student loan debt, Medicare for All, and Green New Deal, that were broadly popular. His top five things he talked about were all things that were broadly popular with the American public, but he put this class language on it. He basically took the vibe and the energy of Occupy and then used it to blast out this messaging around, like, issues and programs that would dramatically improve people's quality of life. What the 2016 Sanders campaign showed the Left—with all the mobilization around it, with all these social movements, because it was much more than what he happened to do—was, "Oh, if you talk about the issues and put forward a program that is already broadly popular with people and they can understand how it will benefit their lives, you can win the support of, like, millions or tens of millions of people. There's no reason for the Left to act in such a self-marginalizing manner. If we just talked to people about the things they most need in their lives and offer them a credible program for how we're going to make their lives better and we have a messenger like Sanders, who's a very compelling figure, emphatically championing these issues, this will resonate with people." That was a huge breakthrough, I mean, more than winning, like, twenty-two primaries and thirteen million votes or whatever. That campaign, I think, just changed people's sense of what was possible.[14]

The Sanders campaigns were not model examples of organizing, however. Van Auken considered organizing for the first campaign, but the neglect of the grassroots bothered her. She decided instead to organize independently, developing a large grassroots group called Team Bernie NY:

> Obama had staff that were very field focused and smart, very well run. Bernie has not had campaigns that were run like that. That's a whole other

thing. Why the Left on a national level can't pull off organizing campaigns in the same way is interesting to me.

Early in the Bernie campaign, I went to this meeting in Manhattan. It was a stupid meeting where everybody was trying to talk high-level strategy. "We should get de Blasio to endorse Bernie!" [conveys pompous speech].

I wasn't interested in this. I was interested in building something with volunteers that was of the caliber of the Obama campaign field stuff that I had done. On the train back to Brooklyn, I ran into someone I'd met at the meeting and who actually lives like two blocks from here [in Sunset Park]. And I was like, "Well, I would love to try and have a meeting where we talk about voter contact and talking to voters and running our own campaign as volunteers." He was like, "Should we have a Brooklyn for Bernie meeting?" This was in July of 2015, a year before the election. We had a Brooklyn for Bernie meeting on his roof, and, like, thirty-five people showed up. I remember being, "This is really cool. There's a lot of energy here. This is a really big organizing opportunity!" So we started Brooklyn for Bernie, and then we changed it to Team Bernie NY because it was the whole city. I met several people who I'm still friends with. We built this wonderful network of brand-new organizers across the city and across the state a little bit too, and I was just like, "Yes, volunteers can do this work in a very disciplined, like, cool way and build community." Then the campaign finally came to New York, a month before the election. It was not a good experience [laughs]. They very much came in and steamrolled a lot of the work that people had been doing and sent people who were kind of assholes to be honest. But we gave them almost fifty thousand IDs that we had collected of supporters across the city.[15]

A Specific Outcome: The Narrative of the "Tax the Rich" Campaign

Sanders's influence on NYC-DSA's vision is illustrated by the first round of the organization's "Tax the Rich" campaign—a major pressure campaign launched in 2019 and reprised in 2023. "Tax the Rich" is a bold Sanders-style slogan that targets income inequality. Both Sanders and Clinton supported a wealth tax in 2016, but the policy was central to

Sanders's platform. He has promoted tax increases on corporations and the wealthy throughout his career.[16]

Many DSA activists read *Outsider in the White House* twenty years later, when momentum for a wealth tax was growing. In New York, socialists and progressives elected during the "Blue Wave" of the 2018 midterms helped Democrats achieve a supermajority. The Working Families Party (WFP) had fought for a wealth tax for decades and led the local Tax the Rich campaign coalition that included DSA. The WFP-backed state senator Jessica Ramos sponsored a bill in May 2020 proposing a tax increase on citizens earning more than $1 million per year. AOC, also backed by WFP, supported the proposal. The campaign ran in late 2020 and early 2021, with DSA conducting its largest phone-banking operation to date: around two hundred thousand phone calls were made to voters, asking them to call their representatives.[17]

The bill was passed during the 2021 budget session, albeit with a reduced funding level. The budgets for public education, the homeless, migrants, and renewable energy were increased by $5 billion. The democratic socialists celebrated this outcome as the end of a decade of austerity policy. The outcome also gave broader public attention to the young New York Left, including critical reactions. *The Wall Street Journal* asked if the bill would make the rich leave the city. Hundreds of readers of *The New York Times* similarly expressed skepticism of the socialists.[18]

The 2022 proposal to run a new version of the campaign shows a well-rehearsed argument against the neoliberal policies of Governor Kathy Hochul and President Joe Biden. It argues that the state budget has a big impact on the living conditions in the city and that the budget negotiations are an opportunity to make ideological differences concrete and relatable: "Do you want private companies competing for every last dollar of profit to provide your child care? To collect your rent? To make sure your kids have clean air to breathe? Would you rather a billionaire's child gets to keep their entire inheritance without paying any taxes, or would you prefer all children in New York get high-quality child care?"[19]

Crisis 2: Trump's Election in 2016

The main catalyst for DSA's growth and transformation was the shock from Trump's election and the frustration with the alternative presented by the Democratic Party. Many young people in the city felt that Trump represented the extreme opposite of their moral values and a setback for decades of hard-won victories for gender equality, racial justice, organized labor, and public education. The fascist aspect was obviously experienced most strongly by people of color. One of them is David Alexis, a poor first-generation Haitian immigrant who would later become a DSA-supported candidate for state senate:

> I'll never forget when Trump won the election. I was driving [Uber] on Election Day. And the amount of fear, trauma, and pain that was there—I picked up Hillary Clinton staffers that day—it was not a pleasant moment. It was a very despondent . . . [looks like he is hurting and takes a deep breath]. This was now the president of the country where I was trying to raise my newborn daughter! I just became a dad. What was I going to do? How am I going to make sure she was safe? What am I going to tell her that this is her president? What can I do? That galvanized me to action.
>
> Trump is part of a racist tradition of vilifying the most vulnerable populations in the community. When he talks about "shithole countries," it hits harder because of that history. He knew what he was doing. It's so polarizing and damaging because he's really jamming into those raw wounds.[20]

An experienced editor of a leftist paper in Brooklyn described the atmosphere in the city as unsettling: "It immediately started to feel like every day is a crisis."[21] The immediate emotional response was followed by an urgent desire to get involved in some form of activism. There was a wave of protest marches in Washington, with satellites held all over the country. One of my informants remembers "an explosion of activism and interest":

> TEFA GALVIS: It took Donald Trump for some people to see what many of us had already seen. Organizing for a social movement under the Obama administration had been extremely hard because people were so complacent. "That's just how life is," you know. When Trump happened, people were like, "Oh, shit! This is what happens when you're just thinking about your own interests!" Many were

shocked by seeing such a person so high up. A lot of the people that started coming out were not directly affected by these issues, but they started to realize that being silent was being part of the problem.

DEVON MCMANUS: I feel like after the Trump inauguration in January 2017, it was the beginning of all the crazy stuff happening. I went to AOC's first-ever canvass back in July 2017. It was through Brand New Congress, not DSA, but she talked about DSA a lot in the early days. That was an interesting race because they started extremely early. They started like a year before. I just went to that canvass. It was very far away, in Astoria. It took me an hour and a half to get there. The Trump presidency was a huge part of it for me. It was like, "Well, I can't do anything about federal politics, but I can contribute to this, and it seems like a way to get really deeply involved." I went to a DSA meeting. Tascha Van Auken was one of the leaders at that point, and I got involved in an electoral campaign in Bay Ridge.

ROBERT WOOD: It was Trump's election that made me really want to go into political activism. You hear that from a lot of people who are members of DSA and people who are in pipeline and climate fights in New York City. After Trump was elected, I realized I had to get involved in something. At first, I was in 350.org, but it was frustrating. I was the one who tried to get them to stop spending time on, like, an antiplastic campaign. Consumer-oriented politics. All this while I'm going to DSA meetings and realizing that these are actually more my people. I eventually left 350 Brooklyn because I wanted to be in an organization that basically puts the politics first. Deep down I've always known that capitalism is the problem, and DSA was framing everything that way.[22]

More than ten thousand millennials joined DSA in the months after Trump's election: David Duhalde has been a member since 2003 and has held different positions in the organization. He reports that the big growth in DSA's membership followed from Trump's victory, not Sanders's primary.[23] Sam Lewis remembers that the formative meetings in the new NYC-DSA were the Brooklyn branch meetings in December 2016 and January 2017.[24]

The young New York socialists could have followed the path of the anti-Trump movement activists and joined the Democratic Party. But their perspective on the party had been transformed by Sanders, and they were disappointed with the party's response to Trump. The NYC-DSA leader Aina Lakha provides insight into this situation:

> So Bernie runs, and a lot of people join DSA, but the bigger bump in DSA happens after Trump's election. It's the dissatisfaction with the establishment of the Democratic Party and with Democratic Party politics and the need to take political action in the face of their obvious failure to be able to defeat Trump. The distinction in the Resistance, broadly defined, might be mapped out onto how you felt about Hillary Clinton and Hillary Clinton losing. There are people in the Resistance whose explanation is, like, "Well, it's the Bernie people who did it." Those people that are obviously beyond the pale for joining DSA, but they're a part of the Resistance. Then the other part is people who are like, "It's because I wasn't involved. More people need to get involved." This very classic sort of liberal idea, like, "We just need to get more involved in this political system, and there needs to be more people who make sure it never happens again. We've got to fight this orange guy or whatever." But the DSA people are like, "Wow, the Democratic establishment totally failed to do the easiest thing in the world, which should have been to defeat Trump. But of course they couldn't do it because they don't have the right economic answer. They represent rich people, not working people, not the majority." And it's that set of ideas, it's that response, that lays the foundation of how people end up joining DSA. But it's very important that thing around Bernie happened first, because it says that there is an alternative.[25]

The Democratic Party has not changed course, but the outcome of the 2024 presidential election triggered concerns among the wider public. When Kamala Harris lost, some mainstream media pundits found that her campaign had neglected ordinary people's economic hardship. A public conversation about revolutionary violence against oligarchy erupted in response to Silicon Valley's support for Trump and the assassination of UnitedHealthcare CEO Brian Thompson. However, there was little public discussion of what an alternative political project might look like and no party to promote such an alternative. That is the most urgent task for a democratic and peaceful solution to the escalating crisis.[26]

4

Anticommunism Kills

In this newly emerging world people want politicians who approach issues without ideological preconceptions.
—Tony Blair and Gerhard Schröder (1998)

Frankly, the lack of political consciousness is exactly what the ruling class of this country wants.
—Bernie Sanders (2016)

In 2022, big oil companies more than doubled their annual profits to a whopping $219 billion. They paid out $110 billion to shareholders. Then, they began rolling back their pledges to scale down, and governments gave them permission to expand production, citing the repercussions of Russia's war on Ukraine. Subsidies increased from $2 billion in 2021 to $7 billion in 2023. Meanwhile, oil companies continue to spend hundreds of millions of dollars every year to promote themselves as climate friendly.[1]

The socialist revival has emerged from moral outrage about basic material issues and eventually also fossil fuel capitalism. Socialists argue that capitalism rewards a small elite for destroying the planet's ecosystems. This argument has a history in the social sciences. In the 1950s, the social democratic economist John Kenneth Galbraith pioneered the argument that private production and consumption were negatively impacting the public commons. Galbraith used air pollution and traffic noise in New York City as key examples, observing that many residents enjoyed the benefits of having a car but that cars were polluting the city. The relationship between capitalism and the environment became central to environmental sociology in the 1970s. Environmentalist critiques of capitalism are disputed, however. Ecological socialism is up against powerful forces, and this situation needs to be addressed before we can understand socialist climate politics in contemporary New York.[2]

This chapter acknowledges how the socialist revival is shaped by a history of struggles with the economic elites and right-wing populism. I argue that the revival is contesting neoliberal capitalism, including the narrative that socialism died in 1989. The ongoing struggles register in competing discourses on "socialism." I first offer socialists' self-definition by asking what democratic socialism means to them and how they relate to this term. I find that the revival is motivated by concrete issues in the present and not the label "socialism." This is not a revival in the sense of a nostalgic desire to return to some idealized socialist past. However, democratic socialism is also being defined externally by opponents in the tradition of Cold War anticommunism. I show how anticommunism feeds symbolic and physical violence against socialists and obstructs democratic conversations around capitalism. The chapter title highlights the perspective that anticommunism thus contributes to ecocide.

The Culture of Contemporary Democratic Socialism

If you tap a friend or Chat GPT for a definition of democratic socialism, you will probably get a textbook account of a particular variation of socialist ideology. One of the problems with such generic definitions is that they abstract ideology from embodied experience and social history. Marx and Engels developed their theory based on observations in London and Manchester in the mid-nineteenth century. The age of industrial capitalism is long gone, and the social structures and technological conditions of production have fundamentally changed. The institutions of the nineteenth-century labor movement have eroded and were not a viable path for young New Yorkers in the 2010s. Many of them were downwardly mobile and learned about socialism in college. Their holistic approach to organizing took inspiration from diverse traditions.

The recent public conversation about socialism was pioneered by the 2016 Bernie Sanders presidential primary campaign. Sanders pioneered a movement focused on the basic material problems that working people are facing today. He was not focused on precise ideological distinctions and appealed to both social democratic and more radical socialist orientations. Ideological purity is not the focus of this movement. Contemporary US democratic socialism is not anti-intellectual, but there is a sense that the seriousness of the situation disallows intellectualizing. Sanders's narrative in 2016 structured the movement discourse: American workers were being run over by a greedy corporate elite. He pointed to a critical decline of real wages coinciding with the increasing concentration of wealth and the influence of the wealthy on politics, including the Democratic Party. He framed these elements in a simple moral narrative and a demand for change: "Enough is enough!" The solutions? A living wage that would create basic dignity and security for all, a single-payer health-care system, and free public education.[3]

Like the wider international democratic socialist movement, Sanders appealed to diverse demographics, including young people in the big cities and workers across the country, thus transforming existing political cleavages (see chapter 2). NYC-DSA is a distinctly young and urban formation in this movement, shaped by the institutions and history of New York City. It has expanded its initial focus on Sanders to situate itself in a broader history of labor movements, New Deal reforms, and the civil rights movement.

When AOC was elected to Congress in 2018, she became a new star of the movement. For years, she and Sanders were the main national voices of the movement. She represented its young urban base, specifically the culture of NYC-DSA. Her status as a movement hero declined in 2024, however, when many people felt that she was aligning herself with the Democratic establishment and not being vocal enough about Gaza.

AOC followed Sanders's concrete approach to ideology and expanded on his conception of socialism. She incorporated racial and environmental justice perspectives and cultivated a more joyful movement culture, embodied in her persona as a young, educated urban Latina with charisma and extraordinary communication skills. In an interview shortly after her election in 2018, she offered the following oft-cited definition of socialism:

> When we talk about the word socialism, I think what it really means is just democratic participation in our economic dignity, and our economic, social, and racial dignity. It is about direct representation and people actually having power and stake over their economic and social wellness, at

the end of the day. To me, what socialism means is to guarantee a basic level of dignity. It's asserting the value of saying that the America we want and the America that we are proud of is one in which all children can access a dignified education. It's one in which no person is too poor to have the medicines they need to live. It's to say that no individual's civil rights are to be violated. And it's also to say that we need to really examine the historical inequities that have created much of the inequalities—both in terms of economics and social and racial justice—because they are intertwined. This idea of, like, race or class is a false choice. . . . There is no other force, there is no other party, there is no other real ideology out there right now that is asserting the minimum elements necessary to lead a dignified American life.[4]

AOC talks from a spirited movement perspective with a universalist appeal to a moral vision of social justice. She is inspired by Martin Luther King Jr.'s ideas and persona as a movement leader. Like King, she has experience in community organizing and challenges the fundamental thinking behind the injustices of powerful institutions in society in a way that is relatable to ordinary people. She brings movement perspectives into critical thinking about institutional politics. I talked to NYC-DSA organizers who volunteered for her first campaign and said that she was "a movement person" from the start, inspiring a feeling of solidarity among volunteers and voicing the interests of ordinary people in her district.

The youthful positive vision of the early AOC that inspired the New York socialist movement can be found in the video for her first campaign in 2018. In this video, she calls out problems but quickly moves on to share her vision for a better future. She never engages in doom talk, and she has distanced herself from sectarianism and ideological purism.

AOC picked up on the spirit that was emerging from movement media in New York City. *Jacobin* magazine led the way with its vibrant visual identity, straightforward and engaging language, and welcoming attitude. *Jacobin* writers introduced socialism to new audiences with books such as *The ABCs of Socialism* and *The Socialist Manifesto*. A more indignant style, defined by irony and sarcasm, can be found in podcasts such as *Chapo Trap House, Citations Needed,* and *The Max and Murphy Show* (renamed *Max Politics* in 2016), all of which have spirited young hosts. Twitter, the platform renamed X in 2023, was a central movement space around 2017. It helped stimulate interest in socialism and DSA. A person named Christian Bowe ran an account named @LarryWebsite with more than ten thousand followers. He kept posting statements such as "DSA is the greatest organization out there!" When new members were asked how they discovered DSA, many mentioned @LarryWebsite. Movement enthusiasm was further boosted by Gallup polls suggesting that socialism had become more popular. However, the big change showed by these polls in the 2010s was growing skepticism about capitalism, not support for socialism.[5]

The new culture of socialism also found a visual expression in NYC-DSA. One of the organization's graphic designers, Ronin Wood, says that NYC-DSA reintroduced the color red and redefined its meaning. He also says that the word "socialism" has not been front and center of its visual identity:

> Before I joined DSA I worked on a congressional campaign for a progressive Democrat. I began noticing that DSA was trying to win people over not just for points and that they were building something with their visuals. I was pretty struck by the red. I avoided red for so long because a teacher in college talked about red as a copout for the designer and because of the connotation of Republicans. But I loved how bold it was. I loved the rose imagery too. The rose imagery has been one of the biggest forms of political identity for me. When I first saw the roses, I thought they were pretty. But then when I started reading about how we don't just deserve bare necessities but a dignified life, I became obsessed with that. It has such a beautiful meaning. A lot of people don't know what DSA stands for, so something that I've been doing is just writing the word "socialist" on things. For some membership drive stuff that I did, for instance, I wrote "democratic socialists" huge on it.[6]

Anticommunism in the Neoliberal Era

The revival of socialism does not make sense from a neoliberal understanding of history. According to this view, socialism was proved wrong when six eastern European communist regimes collapsed in July and December 1989. Francis Fukuyama wrote in *The End of History and the*

Last Man (1992) that the world now considered capitalism inevitable.⁷ The only alternative to capitalism he considered was communism and specifically the totalitarian regimes that long abandoned communist ideology. *The End of History* only mentions social democracy in passing and thus creates a rigid dichotomy between communism and capitalism. On his own terms, Fukuyama was right that communism "has ceased to reflect a dynamic and appealing idea" and that (totalitarian) communists defended a reactionary order like monarchists. The problem is that he implicitly subsumed all forms of socialism and neglected its most popular democratic form in the West. Moreover, Fukuyama rightly observed that postwar prosperity had lifted many people into what Marx called the "realm of freedom." However, one can dispute his claim that there was no longer any serious poverty and that any remaining forms of inequality were largely due to "the natural inequality of talents" and "the economically necessary division of labor." These statements have not aged well.⁸

Fukuyama's book title referenced an old debate in the social sciences, namely, the end of ideology debate of the 1950s and 1960s. The debate began with the argument that the extreme politics of the preceding decades had come to an end. The term "ideology" had different meanings, however. For some, the end of ideology meant the dismissal of totalitarian doctrines, while for others, it meant suspicion of all rigid formulas of ideas and mental illusions clouding human cognition. The phrase "the end of ideology" originates in a debate among leftist intellectuals, with the ex-communist novelist Arthur Koestler arguing in 1955 that the fight against totalitarianism made the terms "socialism" and "capitalism," "Left" and "Right," less relevant. Leftists joining forces to rebut totalitarianism should dismiss these terms, he argued. The public conversation around ideology in the US, however, quickly became dominated by the government's anticommunist campaigns that repressed the Left. This development showed that ideology remained relevant and created more polarized views of the debate. Fukuyama's book followed this pattern by prematurely claiming victory for his own camp and ignoring Koestler's point. Also, state repression against the Left has been revived by Trump.⁹

The events of 1989 did have a major impact on the international political environment. They accelerated a development in traditional leftist parties, with social democratic parties moving right. Instead of providing an alternative to communism and neoliberalism, these parties gave in to pressures from the increasingly powerful capitalist elites and morphed into a form of neoliberalism in disguise. Tony Blair and Gerhard Schröder adopted the concept of "a third way," imagining a path that transcends social democracy and neoliberalism. They stated that markets should be "improved by political action, not hampered by it."[10] The neo-Weberian sociologist Anthony Giddens provided academic support for the Third Way project in the book *The Third Way: The Renewal of Social Democracy* (1998). He argued that socialist ideas and values "remain intrinsic to the good life" but that the economic program of socialism had been discredited. "The economic theory of socialism was always inadequate, underestimating the capacity of capitalism to innovate, adapt and generate increasing productivity."[11] In the spirit of the times, Giddens emphasized economic growth, productivity, and entrepreneurialism.

In the United States, Bill Clinton created a welfare reform that aligned with conservative views and championed free trade agreements, like the North American Free Trade Agreement (NAFTA), against the interests of labor unions. Neoliberalism gained such dominance by the 1990s that a wave of local campaigns for a livable wage was perceived to be controversial. The political scientist Adolph Reed Jr. was alarmed by the opposition to such basic labor justice and the government's involvement in subpoverty-level employment. While the Right conquered the White House, the Left retreated from public life and into English departments.[12]

To repress socialism, neoliberals tapped into the powerful myth of anticommunism in the US. Anticommunism has roots in the religious fanaticism of the Puritan British settlers, who sought to create a purer Christian society. Their influence echoes in today's Christian nationalist movement. State-led anticommunism began in response to the Russian Revolution in 1917 and evolved considerably after World War II. Anticommunism escalated as a politics of national survival in the Cold War with the Soviet Union, the other superpower to emerge from the war. This involved state repression of socialism at home and a military campaign against it abroad, lasting decades. Anticommunism is thus inseparable from Western imperialism.[13]

I witnessed anticommunism firsthand in the field. On one of my lunch breaks at New York University, I went to a diner in the East Village, a largely leftist neighborhood. On this day in April 2022, the diner was packed as usual, and I told the older man sitting next to me that I was researching the socialist movement. His response was not gentle: "Communists! They don't want to work, and they want everything for free!" The word "socialist" activated rage. I learned that many conservative Americans use the word "socialism" as a pejorative and do not feel that this requires justification. When Donald Trump runs out of ideas for bullying his opponents, he simply calls them socialists. He has used the word countless times in his presidential campaigns.[14]

Cold War anticommunism anachronistically frames democratic socialism in the totalitarian Soviet past, while ignoring the social injustices of capitalism. This propaganda linked anyone to the left of the Republican Party to totalitarian communism and viewed them as pathological. It was engineered by strategists hired by Joseph McCarthy, Richard Nixon, and other Republican politicians. Anticommunist propaganda helped them frame mainstream Democrats negatively as "communists," "socialists," "radicals," "subversives," "welfare queens," and "terrorists."

Anticommunism has sedimented at deep, mythical levels in popular culture through Cold War movies showing Russian spies as evil but ultimately inferior monsters. The folklore circulates internationally through platforms such as Netflix. In the dark Nordic winter in 2023, for instance, I watched the series *Young Sheldon* with my then nine-year-old daughter. The protagonist is the boy Sheldon, who lives in conservative small-town Texas. In one episode, he expresses sympathy for communism at school, and the parents panic over the imagined consequences for the family's standing in the community. This is comedy, of course, but it does show the continued power of the myth of communism as a pathology disconnected from its substance. Other examples include the social media crusade by the retired psychology professor Jordan Peterson.[15]

Anticommunism also hits NYC-DSA. The city's real estate industry routinely runs fearmongering campaigns against the organization's candidates. A message of such a campaign in 2022 read, "Warning! Samy Nemir Olivares is approved by the socialists that threaten public safety!" This messaging might have influenced the outcome of this particular race, in which Olivares lost by just two hundred votes.[16]

News media play a key role in boosting anticommunist images of the socialist revival, especially cable networks. Fox News has given voice to bullying and misogyny. A Republican campaign consultant dismissively called AOC "the little girl" on an evening show shortly after her election. A few months later, the network reported on a video of an eight-year-old girl impersonating AOC, portraying her as a self-absorbed child with her own little naïve plan for saving the planet called the Green New Deal. Such blatant gendered bullying trivializes socialism, and in 2019, Fox News was acting like the propaganda arm of the Trump administration. The network boosted the harassment of AOC in the national public sphere. For instance, Republican colleagues in Congress have called her "a f——ing b——h" and "a scared little girl" whose "Green New Deal will destroy the US's oil and gas industry." Julia Salazar told me that AOC always has security guards. This is common for congresspeople, but the anticommunist bullying has put her at greater risk. It is a form of emotional violence that incites physical violence: AOC was one of the main targets on the death list of rioters on January 6, 2021. She was lucky to survive.[17]

Local news media in New York dramatize tensions between neoliberals and socialists in the Democratic Party and offer the socialists little opportunity for self-definition. Said a leading organizer in NYC-DSA, "The media try to make us look small. But we have power. We're getting people elected." *The New York Times* has covered major socialist electoral wins but also allows right-wing pundits to publish cruel misrepresentations.[18]

Socialism in Everyday Climate Conversations

The revival of socialism has spread to everyday life among young people beyond the political field. Socialism is common in popular interpretations of the climate crisis. A poignant example is Sally Rooney's 2021 novel *Beautiful World, Where Are You?* The protagonist, Alice, a novelist from Dublin, Ireland, writes in a letter to her friend about an epiphany she had in the local convenience store. While at the store to buy lunch, Alice realizes that she is entangled in the unsustainable global economy. Interestingly, Rooney presumes that this response is common among urban, college-educated millennials. Basic socialist ecological thinking does not require any introduction.

I was in the local shop today, getting something to eat for lunch, when I suddenly had the strangest sensation—a spontaneous awareness of the unlikeliness of this life. I mean, I thought of all the rest of the human population—most of whom live in what you and I would consider abject poverty—who have never seen or entered such a shop. And this, this, is what all their work sustains! This lifestyle, for people like us! All the various brands of soft drinks in plastic bottles and all the prepackaged lunch deals and confectionary sealed bags and store-baked pastries—this is it, the culmination of all the labour in the world, all the burning of fossil fuels and all the back-breaking work on coffee farms and sugar plantations.

Alice's friend Eileen quickly brings up socialism in her response letter in an equally commonsensical manner. "People think socialism is sustained by force—the forcible expropriation of property—but I wish they would just admit that capitalism is also sustained by exactly the same force the opposite direction. I know you know this."[19] This exchange represents a popular worldview among college-educated millennials. Rooney is one of the many who became sensitized to socialist ecological thinking in humanistic college environments. Downward mobility, inequality, and climate breakdown are common conversation topics, and they are linked. Naomi Klein's 2014 *This Changes Everything: Capitalism vs. Climate* provides the template for Alice's epiphany and her moral dilemma about writing novels when the world is on fire. Alice finds hope in the belief that the love between humans will eventually save humanity. *Beautiful World, Where Are You?* does not venture into political activism and organizing, however.

The circulation of ecological socialism as a folk repertoire can also be identified in the work of another millennial author. Jenny Odell's 2019 *How to Do Nothing: Resisting the Attention Economy* is an existential exploration of coping with multiple crises. It focuses on capitalism's demands on the individual for productivity and self-promotion. Odell finds that capitalism thus distracts people from important aspects of human life and from nature. For Odell, the story begins with a collective existential crisis following Trump's election in 2016. Instead of becoming involved with politics, she turned inward. She withdrew from the noise. Odell is an artist based in the San Francisco Bay Area, and she started going daily to a nearby rose garden "to do nothing." From there, a new sense of being emerged with a perspective on the history of her local environment. This includes the story of trees that survived for over one hundred years by being too weird or useless for industrial exploitation. Odell argues that shifting focus away from the individualist attention economy is necessary for tuning into the public, physical realm. This faith in an embodied mode of existence also appears when she is teaching her regular class on digital design and feels compelled to ask, "What does it mean to construct digital worlds when the actual world is crumbling before our eyes?"[20] Odell explores the idea of resistance-in-place, referring to the individual's negotiation of the conditions of life. There is a sense that the world will change for the better if more people follow her example.

Whereas Rooney's protagonist provides a standard leftist interpretation of climate change, Odell is starting to live differently. But it is still about individual, not collective, action. There is no plan for political change. Odell describes resistance-in-place as hard work for the individual.

Why not work collectively to change the conditions?

5

A Departure from the Democratic Party in Brooklyn

Whitney, a Movement Person

In my first week of fieldwork in New York City in March 2022, I stumbled into a petitioner on a crowded street corner in Flatbush, Brooklyn. I had conducted research in the city before but never systematically explored Brooklyn beyond Williamsburg. This time I landed in Brooklyn after eight years away from the city.

I had rented a place in Park Slope with the expectation of conducting fieldwork there for a somewhat general book on media culture. My plans were scuttled by the ongoing lockdown, and I began searching for a relevant community to engage with. For a few days, I talked to all kinds of people on the streets of Park Slope. I experienced a high level of civic privatism among locals, no doubt exacerbated by the pandemic. People did not have much time or interest in improvised conversations with this stranger. I like Park Slope, but it felt like an affluent enclave, shielded off from major societal tensions.

I therefore began exploring other neighborhoods, starting with nearby Flatbush. I biked through Prospect Park, a sprawling century-old park designed by the same architects as Central Park and boasting similar wide-open spaces, lakes, extensive bike paths, and other recreational facilities. The park's undulating meadows, curving pathways, and rich woodlands mimic pastoral nature. The park provides a passage from the affluent and predominantly white Park Slope to the lower-income and more diverse Flatbush. I entered the park by the Brooklyn Public Library and the area of multimillion-dollar brownstones on Prospect Park West. On this Sunday morning, as on most days, many locals were walking their dogs, exercising, and going to playgrounds with their kids. Prospect Park is more of a local neighborhood park than is Central Park, which is a hypermediated and more touristic space. I confess to being drawn to the image of Brooklyn as an authentic alternative, and I found a particular form of authenticity in a political grassroots movement that is critical of mass media and corporate capitalism.

I exited the park to Flatbush Avenue, with its many dollar stores, cheap fast-food eateries, and churches. Flatbush Avenue was distinctly different from Park Slope and had generally not been gentrified. On my first day on Flatbush Avenue, many people were unfamiliar to me, and I did not know how to approach them. I felt like an outsider. I had to make efforts to develop conversations. By the end of my fieldwork two and a half months later, I felt that I could talk to anyone in Flatbush. I learned to do so from volunteering as a canvasser for an electoral campaign and integrating neighborhood ethnography into my canvasses. I engaged in substantive conversations with roughly two hundred constituents during my canvasses and interacted with locals on my routes, including small business owners, delivery drivers, pastors, drug dealers, and people experiencing homelessness. But on this first day in Flatbush, I was a fumbling, white college professor from a faraway country. It was easier for me to "fumble" because as a foreigner, I was less entangled in local histories of social injustice. I was not an "outsider within."[1]

An elderly Haitian immigrant was sitting on a chair outside a clothing store. She was watching something on her phone. When our eyes met, I asked, "Oh, you have a smartphone. Do you enjoy using it?" "I don't have a smartphone. I have an iPhone," she replied. I was baffled and asked what she preferred to watch on her phone. She mostly used the phone to watch Haitian TV shows. She had a second phone under the federal Lifeline program for the poor. Further down the street, a seventy-year-old man was waiting for a friend. He had lived in the neighborhood for thirty years and loved the community and the churches. He was on his way to church. This was a Sunday, after all. He detected my curiosity and encouraged me to go a bit farther down Flatbush Avenue and into the Church Avenue area. "That area is different. It'll be interesting to you. You should go there!" On the corners of Church and Ocean Avenues were diverse groups of people. People from the surrounding high-rises, some of them social housing, and prewar four-story buildings were shopping for groceries, smoking, getting a haircut, and buying fish, and a few were buying drugs. There were mostly people of color, and many of the shops were Afro-Caribbean.

There was also a person on the corner talking to passersby and handing out fliers. Her messaging intrigued me:

WHITNEY WITTHAUS: Hi! Do you think we should tax the rich?
[Traffic light changes, cars start running.]
PASSERBY 1: Uhm ... [looking puzzled, walks away].
WITTHAUS: [To another passerby] Hi! Do you support universal health care?
PASSERBY 2: Hey [nods and continues walking].
WITTHAUS: I'm Whitney. [A car honks and speeds up.]
PASSERBY 3: [Ignores Witthaus]
[Two dogs are barking. The traffic light changes again.]
WITTHAUS: Hi, I'm Whitney! [passerby makes eye contact and stops] I'm collecting signatures to get David Alexis on the ballot for the state senate race in June.
PASSERBY 4: Yeah?
WITTHAUS: I'm from the Democratic Socialists of America. We are fighting for universal health care, free public education, tenant rights, and a livable climate future. David's campaign promotes our proposed legislation for a socially just transformation of the energy sector in New York State.
PASSERBY 4: Yeah, I'm really concerned about climate change.
[The sidewalk gets more crowded.]
WITTHAUS: Right, and New York hasn't passed any climate legislation in three years. The current senator for this district, Kevin Parker, is allowing this to happen. He's chair of the Energy Committee, and he is funded by the fossil fuel industry! You can help replace him with someone who will actually fight for climate and working people in Flatbush: David Alexis is a working-class father, a democratic socialist, and he's not taking corporate money.
[Three drug dealers nearby are giving me a skeptical look.]
PASSERBY 4: Sounds good ... [signs the petition and quickly moves on]
[At this point, I've been standing near Witthaus for a few minutes, while she has been talking to pedestrians.]
AUTHOR: Hi, excuse me, I overheard the conversation. Can you tell me more about what you are doing and this organization that you mentioned?

WITTHAUS: Yeah, Democratic Socialists of America. We're a movement. We're now petitioning around Flatbush to get Davis Alexis on the ballot. The campaign is just starting. Do you want to sign the petition?

AUTHOR: I'm not a citizen, but I'm doing research for a book and would like to know more about your organization.

WITTHAUS: You're welcome to volunteer with us. We have a social event after every canvass where you can meet other volunteers. How about next weekend?

AUTHOR: Uhm . . .

WITTHAUS: It's really easy and fun! Just go on the Alexis campaign website and sign up. Where are you from?

AUTHOR: I'm from Denmark.

WITTHAUS: Oh, Denmark! I'm a schoolteacher, and I love teaching welfare society and social democracy in Denmark and the Nordic countries. Universal health care and free education! You know, it's really interesting. . . . [Witthaus explains how she teaches Nordic welfare society]

Whitney Witthaus's message and her interactions with pedestrians triggered my interest in DSA. Her clear presentation of a substantive political project focused on big and urgent issues, and her capable interactions were like nothing I had experienced before from a political campaigner. Witthaus understood the core political ideas of the campaign, its broader political context, and its relevance to life in the neighborhood. She was not simply reproducing a campaign slogan or praising the personality of a candidate. I immediately recognized Witthaus's individual talent, but I would soon learn that her political agency and orientation in the world had been shaped by the movement culture in NYC-DSA.

Witthaus's background is typical of NYC-DSA. She is in her late twenties, white, and part of the downwardly mobile middle class. Witthaus studied gender and women's studies at the University of California at Berkeley in San Francisco and became involved with YDSA. After moving to New York, she joined the local labor branch of DSA and its rank-and-file program, in which members choose a place to work with the aim of unionizing colleagues. Like most participants in this program,

Witthaus became a public schoolteacher. The teachers' unions are the largest single segment of unionized workers remaining in the country, but they have been weakened severely by the pro-charter-school agenda of the Democratic Party. The city's Democratic mayor, Eric Adams, has slashed school budgets.[2]

I began participating in the David Alexis campaign to learn more about NYC-DSA, and this led to a transformation in my political consciousness and understanding of political culture. NYC-DSA demonstrated a new approach to climate organizing and inspired me to concentrate on this area. I soon learned that the Alexis campaign doubled as a movement tactic in a multiyear campaign for climate legislation and found the integration of movement and institutional politics interesting. I had never fully engaged in politics before, so my understanding of political culture was limited and largely based on media representations. I therefore had a steep learning curve ahead of me. I spent years processing the field experience and conducting follow-up research.

How can NYC-DSA be situated in the political field in Flatbush and Brooklyn more broadly? Leaders of the traditional Democratic Party youth clubs felt ambiguous about NYC-DSA. They viewed the independence and vibrant movement culture of NYC-DSA with some envy. While they were struggling to be heard within the party organization, their peers in DSA were confidently setting agendas and organizing electoral campaigns with thousands of engaged volunteers. But there was also a sense that DSA was not quite playing by the rules and had a romantic self-image of being a holy savior against evil forces. A leader of one club described DSA as "too idealistic," implying a preoccupation with ideology. Another speculated that the organization was recruiting working-class people of color to appear authentic. One can also see this as a way of promoting cultural diversity in political life.

The crisis between the Democratic clubs and the party leadership in Brooklyn was severe. The clubs were fighting for a more democratic and progressive party organization, like they always had, but the terms of the struggle had changed. They had become paralyzed by a new level of repression from above. Leaders of New King Democrats and Brooklyn Young Democrats reported deep frustrations with the person who had served as chair of the Kings County Democratic Committee since 2020, Rodneyse Bichotte Hermelyn. They claimed that the party was now run like a syndicate; it was blocking clubs from having influence and refusing to communicate with them. One club leader described how the conflict was hurting participation. Nonetheless, the club leaders wanted to improve the party organization, not abandon it like DSA.³

How did NYC-DSA members view the situation? They agreed that their organization was a sort of rebel alternative to the Democratic Party, but their self-definition adds nuance.

Bigger than Bernie: An Insider Account of DSA

Before entering the field on the ground, let us look at how insiders have mapped the national DSA organization in political history. The best source for doing so is Meagan Day and Micah Uetricht's book *Bigger than Bernie: How We Can Win Democratic Socialism in Our Time* (2020). The authors had been active in NYC-DSA for years and chronicled the scene for *Jacobin*. Some aspects of their account will be familiar from chapter 4.

A central theme in the book is that DSA's main opportunity in the late 2010s was to leverage the momentum generated by the first Sanders campaign. The authors begin by framing the national situation in terms of a drama with Sanders speaking up at a time of crises and spearheading an insurgence in the Democratic Party. Sanders addressed the injustices of capitalism and the moral failures of the Democratic Party. The overarching story here is a moral battle led by the heroic Sanders. The party pushed him out of the primaries in 2016 and 2020 by maneuvering in favor of candidates preferred by corporate donors. We learn that Sanders made socialist class rhetoric seem commonsensical in a presidential primary campaign and that his two presidential primaries created "a kind of ad hoc left-wing political party." The authors argue that his primaries offered a political platform, united disparate elements of the Left, and raised expectations for political gains among millions of working-class people. The campaigns "reshaped the terrain of American politics and produced an opening for the American Left."⁴

Day and Uetricht argue that DSA has gained importance because of its electoral work. It is the only organization that systematically develops grassroots electoral campaigns on the Left. The authors also suggest that DSA's growth happened because it was the only socialist organization to fully declare its support for Sanders from the start and adopt his class rhetoric. AOC's and Julia Salazar's victories in 2018 boosted the organization, and its relatively small number of electoral seats, most of them in state and city governments, have symbolic importance: Despite only having around 150 electeds nationally—the number reached 212 in 2025—DSA's victories show that socialists can challenge Democratic incumbents with corporate sponsors and disrupt the status quo. A few victories can change the dynamics of the field and shift the balance of power in legislative decisions. Additionally, we learn that DSA also grew because it has low barriers to entry, has a nonsectarian culture, and runs campaigns that inspire and move people into action. Its demographic consists of a downwardly mobile middle class.[5]

Day and Uetricht then caution that the entry into electoral politics carries the risk of being co-opted by the Democratic Party. They describe DSA's so-called dirty break strategy, in which socialists organize independently but use the party's name and ballot. We are told that there is consensus in DSA that a "clean break" of creating a third party, with the expectation that if you "build the infrastructure," "voters will come," has proven unsuccessful. A clean break would result in obscurity and bypass the opportunity to challenge the definition of being a Democrat.[6]

Bigger than Bernie does not discuss the party dimension of DSA. The concept of a party is thin and vague, and there is no concept of a movement party in the book. The conversation about the party dimension had not developed far when the book was published in 2020.

I was initially puzzled by the excitement about the socialist movement, because its electoral achievements are limited. If Sanders had won the presidential election, sure, that would have been something, but he lost two primaries. My explanation is that the political situation in the country was so dire that even the small achievements of the socialist movement and the emergence of an organized Left were major events.

This sense that the country is in such a deep political crisis that even small victories are celebrated for their symbolic value can also be found in another book of 2020. The premise of McAlevey's *A Collective Bargain: Unions, Organizing, and the Fight for Democracy* is that the United States is "stuck with a high court that will rule against workers and the planet for another thirty to forty years." McAlevey is convinced that lawsuits and legal tactics, the modus operandi of progressives since the

1970s, no longer work. She too identifies economic inequality as the root cause of the major problems in society, but she does not have hope for the political system. The system is too dominated by the economic elites, she argues. McAlevey cites research showing that "ordinary citizens have virtually no influence over what their government does in the United States." McAlevey believes that unions have the greatest potential for changing the situation. I agree with much of her analysis but will not rule out the potential of a leftist movement party. My key point here is simply that the bleak state of leftist organizing at the national level defines the conditions under which the new NYC-DSA emerged.[7]

The Political Field in Brooklyn

The new NYC-DSA intervened in the political field in Brooklyn and represents a departure from the traditional culture of the Democratic Party. This is significant because it marks a transformation in one of the largest Democratic strongholds in the country for over a century, which is also home to a large young population. Brooklyn, which is coextensive with Kings County, had 1,149,287 registered Democrats and 141,652 Republicans in 2023. It is the most populous borough in New York City, with approximately eight hundred thousand people between the ages of eighteen and forty.[8] NYC-DSA follows the pattern of young reformist movements in the party but expands beyond their framework. It creates a new structure outside the party and articulates broader changes in social life.

I detail this argument in the following analysis of NYC-DSA's place in the longer history of the Democratic Party in Brooklyn. I focus on the party's organizational history. My main source for this history is Jerry Krase and Charles LaCerra's 1991 book *Ethnicity and Machine Politics*, about the Madison Club. This club was the most powerful Democratic political organization in Brooklyn from the 1910s up to the 1970s. The book presents itself as an alternative to the dominant focus on celebrities and big national events in political research, and the book is rare for its empirical knowledge of local political life among ordinary people.

A Departure from Neighborhood "Machine" Culture

NYC-DSA has only recently entered the political field in Brooklyn that has evolved for more than a hundred years. Compared to the Democratic Party organization, NYC-DSA is almost exclusively operating at the level of state government. It has little involvement in the city council or federal government. The organization is primarily motivated by a response to events in the national political environment, not from traditional neighborhood institutions. It evolved as an alternative to the traditional culture of the Democratic Party and its "machine politics."

What is "machine politics"? Long into the twentieth century, urban political life was organized around political "machines," defined by "bosses" of political clubs who exploited disadvantaged immigrant populations with limited welfare services. The most powerful political organization in Brooklyn from the 1910s until the 1970s was the Madison Club, created just seven years after Brooklyn became part of New York City in 1898. It was more powerful than the Kings County Democratic Club. The Madison Club was centrally located in Crown Heights and achieved absolute control of the city and all elections through satellite clubs. Its leader, John McCooey, created the club as a way of establishing a political base for himself. He modeled it on the Tammany Hall club in Manhattan, which had pioneered machine politics in the nineteenth century.[9]

The word "machine" has historically been attached to ethnically marked groups—Irish, Jews, and Italians. The machine is defined by a powerful club with an autocratic "boss" leader and satellite clubs that serve as the informal arm and mobilizing tool for the party. The boss is elected by elite networks in the party, not voters. The boss exploits politics for personal benefit. McCooey was the chair of the Kings County Democratic Committee for more than twenty years, but he was never elected to public office. Clubs were apolitical in that they did not articulate a political program. Idealism was anathema to machines. The clubs used their power in informal neighborhood economies to help secure office for political leaders.[10] They mediated between illegitimate and legitimate organizations, with patronage as a key structuring principle. Clubs provided constituents with food, coal, jobs, political influence, and legal protection in return for political loyalty. Franklin Roosevelt, who rose through the ranks of the Democratic Party in New York State, became one of the most prominent critics of the urban political machines. He argued that they should be eliminated because they stood in the way of good government.[11]

The decline of traditional machine culture in the late twentieth century is important to understanding NYC-DSA. The most important reason was that clubs lost control over resources as the government expanded and professionalized welfare services. Formal networks replaced informal ones. This process began with the New Deal reforms in the 1930s and evolved further in the 1950s. Clubs were also subjected to stricter government regulations, and many constituents emerged from poverty. Yet another factor was the limited adaptation capacity of clubs. Their networks were aging and adapting poorly to complex cultural change in the city. Clubs became places of nostalgic conversations. The county machine became more distant from local communities because it became too focused on its own interests and relied more on media and political experts. It ran campaigns with large advertising budgets and relied more on corporate interest groups. The decline of the machine shows how power gravitated away from local personal encounters to more geographically dispersed mediated relations and to more impersonal bureaucratic and corporate structures. This shift happened as part of broader institutional, demographic, and technological changes. The organizational culture of the Democratic Party was now a far cry from the days when "boss" McCooey personally met with as many as two hundred constituents every day in his clubhouse in Crown Heights, settling disputes and handing out gifts and favors with an understanding that the person receiving help would reciprocate.[12]

In the language of Anthony Giddens, the preceding history illustrates the compression of time and space in modern societies. In this process, social relations are lifted out of their local contexts and reembedded into translocal networks and economies based on new technologies. The compression of time and space in Brooklyn has eroded the traditional culture in the Democratic Party and stimulated the emergence of NYC-DSA from other flows of information and power. The new NYC-DSA emerged quickly from media communications about national political events that created a sense of new opportunities in political life. I thus draw from the sociology of modernity to explain what the young traditional Democrats were struggling to articulate, namely, the logic behind NYC-DSA's detachment from the party's traditional culture. The young traditional Democrats felt that the new NYC-DSA merely lacked local grounding. They did not recognize how the organization rearticulates social and institutional relations in the wider geographic contexts of democratic socialism, neoliberal capitalism, and climate change. Rhoda Jacobs, a state assembly member for Flatbush from 1978 to 2014, feels that DSA is a kind of strange and fleeting mobile force. Jacobs is not far from DSA politically and has similarly operated outside the sphere of the Democratic establishment. But she has a longer relationship with the neighborhood than do the young DSA members, who have settled more recently in Flatbush. "Some of these progressives are coming from other neighborhoods," she says, suggesting that they are lacking experience in this neighborhood.[13]

While NYC-DSA's presence in Flatbush was fairly recent, dating to 2022, the organization had done base building for years in ways that had escaped the attention of locals in the Democratic organization. When I met Jacobs, NYC-DSA had just launched its first electoral campaign in the neighborhood, the David Alexis campaign for state senate. The campaign office had just opened, and people in the organization came from around the city to help get the project off the ground. But the campaign was the outcome of the organizing efforts of the Ecosocialist Working Group (ESWG) in the neighborhood since 2019. The ESWG was primarying the incumbent senator, Kevin Parker, because he had been delaying climate legislation for years. People in the Democratic Party organization did not talk about NYC-DSA's base building or about climate change.[14]

The legacy of machine culture persists in the Democratic Party and is a source of moral indignation among socialists and progressive reformists alike. The Democratic mayors Bill de Blasio and Eric Adams have been bribed and have personal ties with illegitimate organizations, and the party's county boss, Rodneyse Bichotte Hermelyn, has forged signatures in petitions, while challenging the petitions of opponents, including DSA-backed candidates.[15] The neo-machine culture extends to the state level, and the socialists have reacted more directly to this. AOC spoke for socialists and progressives when she criticized the state leadership after the 2022 midterms:

> Much of the political machinery that [former New York Governor Andrew Cuomo] put in place is still there. And this is a machinery that is disorganized, it is sycophantic. It relies on lobbyists and big money. And it really undercuts the ability for there to be affirming grassroots and state-level organizing across the state. And so when that languishes and there's very little organizing happening, yeah, I mean, basically, you're leaving a void for Republicans to walk into. And so I actually think a lot of these Republican games aren't necessarily as strong as they may seem, I think it's really from an absence. And it's a testament to the corruption that has been allowed to continue in the New York State Democratic Party.
>
> A lot of this was really about these calcified political machines being asleep at the wheel, and there being a complete lack of desire to hold any of it accountable. . . . I've been in Congress for four years, I have never had a conversation with the New York State Democratic Party chair ever. In fact, he's done nothing but attack progressive Democrats all across the state.[16]

This statement helps us understand NYC-DSA as a democratic and moral alternative to the Democratic Party machine culture. The young progressives in the party felt abandoned by the party, but they remained loyal to the party. Their peers in NYC-DSA chose to organize independently of the party.[17]

A Departure from Jewish and Black Elites

Race, ethnicity, religion, and class are major structuring forces in the US metropolis. A hierarchy of difference is spatially articulated in the demographics across neighborhoods. Historically, the metropolis has white, affluent neighborhoods in the center, surrounded by neighborhoods of low-income Black and immigrant communities. The emergence of NYC-DSA in Brooklyn represents both the expansion of the affluent center through gentrification and the entry of immigrant populations into political life. The latter is particularly relevant to Flatbush and the David Alexis campaign.

Brooklyn developed into the large county that it is today from mass immigration from Italy and eastern Europe from 1890 to 1930. In those four decades, Brooklyn's population grew from 599,495 to 2,560,560.

Many European Jews settled on the Lower East Side. In subsequent decades, many settled in Brooklyn or moved to new middle-class areas in Central Brooklyn.

The waves of immigration register in the Democratic Party. The first machine boss (John McCooey, 1905–1934) was Irish; the second and third (Irwin and Stanley Steingut, 1934–1978) were Jewish. About half of the elected officials in the New York State government were Brooklyn Jews from 1950 to 1968. The civil rights movement in the 1960s stimulated the emergence of Black, Afro-Caribbean, and Hispanic leaders in the party. These changes in leadership also reflect the transformations in immigrant communities described by Daniel Moynihan and Nathan Glazer in their 1963 book *Beyond the Melting Pot: The Negroes, Puerto Ricans, Jews, Italians, and Irish of New York*. In this sequential theory of urban change, ethnic immigrant groups first struggle for resources and recognition, often against each other, before defining themselves as fully American and less in relation to their immigrant origins once they have reached a certain status in society.[18]

How does this history shape political life in Flatbush today? The Jewish community remained influential into the early 2000s but has gradually given way to the Haitian community that began arriving in the 1960s. Haitians fled François "Papa Doc" Duvalier's regime, and many settled in Flatbush, where cheap housing was available. A central part of Flatbush and East Flatbush developed into a "Little Haiti." Of the neighborhood's 156,159 residents, 41.6 percent are foreign-born, mostly Haitians and Jamaicans; 15.9 percent live below the poverty line. The median household income is less than half of that in neighboring Park Slope.[19]

Many of the Haitian storefronts from the 1960s and 1970s are still standing today, and there are signs of Haitian heritage everywhere in Little Haiti, in music, food, and fashion. The Haitian Day Parade and Festival is a source of pride and a promotional platform for local politicians. Most of the performers are young and born in Brooklyn. Most of the music heard from the cars of young Haitians today, however, is English-language rap music. Fast-food chain stores and gourmet cafés are popping up.[20]

There are considerable demographic variations within Flatbush, and it is not a singular coherent entity. Rhoda Jacobs has lived in the neighborhood since the 1960s and says that the real estate industry adopted "Flatbush" as an umbrella term for branding purposes. Jacobs also remembers that Flatbush was "bombarded" by redlining in the 1970s, with banks and insurance companies refusing to offer services to Black residents in certain areas and thus deciding where whites and Blacks live. The areas west of Flatbush Avenue (South Prospect Park, Ditmas Park, and Midwood) are relatively more affluent, and the population is 40 percent white; while those east of Flatbush Avenue (East Flatbush, including Little Haiti) are less affluent and 4.1 percent white.

The social geography has evolved historically. When Flatbush was still countryside, there was Ditmas Park, an affluent suburban village, where the powerful people were slave owners. A mass influx of people from Manhattan began after Brooklyn became part of New York City in 1898 and was connected through new roads and subway lines. Large apartment buildings were constructed in South Prospect Park, some of which are still rent-controlled. Today, Ditmas Park is still the most affluent area, with many progressive white baby boomers and a median household income of $153,653. There are blocks nearby with a median household income of $31,611. Midwood to the south was a mix of middle and working class into the 1960s; but it has since become more affluent and conservative, and its Jewish population has become more Orthodox. East Flatbush became the primary destination for Caribbean immigrants in the 1960s and is marked by urban decline, with a large number of empty storefronts, although there are signs of emerging gentrification. The eastern part of East Flatbush has a high crime rate and borders the low-income and predominantly Black areas of Brownsville and East New York.[21]

Where do you think the David Alexis campaign focused its efforts? We canvassed mostly in the large rent-controlled apartment buildings in South Prospect Park, which has middle- and working-class demographics. We did not focus on areas with a median household income below $45,000, such as Little Haiti. This is not to say that the campaign concentrated on the core NYC-DSA demographic. We canvassed a lot in apartment buildings on Ocean Park Avenue where working-class residents had difficulty covering basic expenses, and many of them were people of color. Some of the buildings had a desolate and eerie atmosphere.[22]

NYC-DSA is also not part of the local elite networks in political life. The political elite in Flatbush is defined by middle-class Blacks of Haitian descent, who have close and long-standing personal ties with leaders of local schools, churches, and other powerful neighborhood institutions. The "bosses" are Rodneyse Bichotte Hermelyn and Kevin Parker. They have ties with party leaders at higher levels, such as New York City Mayor Eric Adams, New York Attorney General Letitia James, US Senator Kirsten Gillibrand, and US Representative Yvette Clarke. They also have ties with local leaders in city council and with leaders of Haitian churches such as Gil Monrose, who is an influential figure in Brooklyn religious politics. The Democratic elite is thus baked into a community power structure that has developed for decades, a gated ecosystem across all levels, from city council to Congress. Bichotte Hermelyn and Parker are affluent and have advanced degrees in business management and political science, respectively.[23]

The Alexis campaign represents a departure from this circuit of the tradition-based Democratic elite. It was the organization's first electoral campaign in Flatbush and an explicit attempt at ousting Parker. NYC-DSA had won over Democratic establishment incumbents in North and Central Brooklyn in 2018 and 2020 and adopted the same approach to expanding southward to Flatbush. A growing number of active members were living in Flatbush, and the Ecosocialist Working Group had campaigned there since 2019, as noted earlier.

NYC-DSA does not have ties to elite networks in Flatbush and contested its power by primarying Kevin Parker with David Alexis, a young working-class candidate without elite ties, except perhaps from an endorsement by AOC. Alexis is in his early thirties, has a bachelor's degree, and works as a home carer and taxi driver. He got involved in NYC-DSA's health-care and climate working groups because he felt they were urgently relevant for improving his basic material needs. Alexis lives paycheck to paycheck and is the primary breadwinner of his family. His wife has sickle-cell disease and is unable to work for that reason. He has two small children. The family lives in a small apartment in lower-income East Flatbush. Alexis is a generous, kind, and moral person, and he is passionate about politics. During his long hours of driving, he has listened to audiobooks and podcasts about socialism, the civil rights movement, and African revolutions, among other things.

However, Alexis is aware that his lack of elite social capital is a barrier to entering the political class. He does not feel entitled to define the world for others like others at NYC-DSA who have an elite college degree, but he has come far through strong personal relations with his comrades in NYC-DSA, as well as with his brother.

> I don't have the pedigree, you know, the experience of working on policy development. I feel like I'm not well-spoken enough. I honestly thought that I would be like the guy on the campaign, you know, knocking doors, doing all the unsexy stuff.
>
> My brother is a big part of who I am. He was part of DSA before I was, okay. He always challenged me because once he realized that I was starting to go on this journey, he took a lot of time. A lot of the initial political conversations I had with him, we'd spend hours on end talking about politics, and he got me into Left podcasts—*Citations Needed, Chapo Trap House, Jacobin*, you name it, all the different major Left voices. And as an Uber driver, you know, there's plenty of empty space and time, so I'd listen for hours, audiobooks, these types of things while I was driving. I've always been interested in history, but I was introduced to this political economy perspective that is lacking in the American school system. That's when I began to understand capitalism and its problems. As a younger man, I thought that the instability in my family was something I had to fix through, you know, the prosperity gospel—these ideas are endemic in the Black community because we've been divested so much—so I thought I just needed to work or start a business.[24]

Alexis is being socialized into NYC-DSA. He is beginning to understand his struggles with basic material needs in broader terms of structural racism. In his speeches during the campaign, Alexis often emphasized his family's situation and struggled to build a case at the level of state politics. He was learning the narrative of democratic socialism and trying to incorporate it into his challenging work life. Unlike most leaders in NYC-DSA, however, he has a strong understanding of racial justice from a working-class perspective, and he had practical labor organizing experience.

My political education was atypical. It didn't happen in the academy at all. I have a bachelor's degree, but I don't have the same formal education, like a lot of other people do. While they went to college, my political education primarily happened through podcasts, books, conversations, and lived experiences. One of the cofounders of the Drivers Coop has been in the labor movement since he was a teenager. When he was a teenager, he organized fast-food workers. We didn't organize from theory or history. We focused on the tactics of organizing, you know, raising consciousness in the workplace hands on, getting people together, and having tough discussions.[25]

Over the course of the campaign, Alexis painfully learned that the party elite was controlling neighborhood institutions through personal ties with community leaders. Prominent church leaders loyal to Kevin Parker denied Alexis the opportunity to meet their congregations. Alexis spoke at fewer and less influential churches, and the national leftist media outlets did not have a strong presence in the neighborhood. AOC endorsed Alexis, but she did not campaign with him in the neighborhood. His campaign relied almost solely on movement canvassers connecting directly with individual voters at the door and collecting microdonations.

Alexis lost. He got 7,047 votes, while the nineteen-year incumbent Parker got 8,543, but Alexis's campaign built enough pressure that Parker got behind NYC-DSA's proposed climate bill (see chapter 9). What can we make of the loss? The folklore in NYC-DSA is that Alexis lost because of a spoiler in the race, Kaegan Mays-Williams. She got 3,034 votes.[26] I think there are two important lessons from this race: The first is that the Democratic Party's organization still has much power in Flatbush through deeply embedded cultural relations. NYC-DSA was unable to overcome this power structure and did not confront it strategically in the campaign's media communications.

The other lesson is that NYC-DSA made the somewhat idealistic choice of recruiting a candidate with the right values but not enough social capital to have a fair chance in the political field. The campaign outcome cannot be evaluated solely on the basis of the candidate's individual qualities, not least because most of the campaigning was done by a large number of canvassers. However, the candidate's relationships and public image matter in electoral races, and my interviews suggest that the campaign recruitment committee could have adopted a more rigorous approach to the candidate's qualifications, resources, and networks in the political field. The people who recruited Alexis talked about how he fit in but not about his leadership qualities or his limited resources. The race would have been fairer if Alexis had not had to work full-time throughout the entire campaign. Alexis could have gone further if he had time to do more public appearances, do more canvasses, and communicate more and better on social media.[27]

A New Actor in the Trump-Era Landscape

The structure of a dominant centrist faction and a small progressive faction in the Democratic Party evolved after Trump's election. But first, let us consider the experience of Rhoda Jacobs to further nuance the cultural transformation in the political field. Jacobs was the sole progressive elected official in Flatbush for thirty years. She got into politics through a national movement in the 1960s, as most NYC-DSA members did in the 2010s. This was the anti–Vietnam War movement. Jacobs also opposed the redlining of Blacks in the neighborhood and wanted to democratize the party. Her values were at odds with the machine. Consequently, the machine tried to get rid of her.

> I was elected in 1978 and had no concept of breaking into the machine, "the organization," as we called it. They blocked me out. But I learned that you needed three *I*'s in Brooklyn to get you on the ballot and win: Italy, Ireland, and Israel. During my first term, the demographics changed radically from white ethnics to Black Caribbeans. I had a primary every two years, largely because I had a community that was perhaps 12 percent white. I had to work very hard in terms of constituent services and really beyond legislation to provide service for the community. I was a white Jewish woman in an essentially Black community. There was a point at which "the organization," as we called it, was perfectly happy to support somebody running against me, and so we had a showdown, and they stopped. But they never did anything for me. The trade-off for me to become a sycophant was not worth it.

So this was basically, you know, children play alongside each other but not with each other? Parallel play.²⁸

Jacobs and the younger generation in NYC-DSA have more in common than either side acknowledges, mostly because they operate in different social circles. For both, the frustration with the party's centrist faction and limited internal democracy has shaped their trajectories as independents.

When Trump won, the relationship between centrists and progressives was framed by new tactical considerations, and the progressive side became more powerful. The centrist faction felt that progressives were hurting the party's popularity and made it more difficult to win over Republicans. "Everyone unite around the centrists, so we can win over Trump," read the message. But Trump's victory led to this surge of a larger and younger new Left with the Blue Wave, and the progressives became a larger faction in New York that now works in alliance with NYC-DSA, working together to move the needle to the left. This has amplified tensions in the party.

A politically active Democrat for decades, Kathryn Krase articulated this complex situation for me. She is a progressive but decided to support a centrist for tactical reasons and feels deeply ambivalent about her decision. Kathryn grew up in a Park Slope brownstone as the daughter of the sociologist and progressive activist and organizer Jerry Krase, who has a working-class background. The childhood home was a hub of progressives and visited by Democrats of all ranks. The family has close ties with Rhoda Jacobs. Kathryn first got involved in a political campaign for a progressive candidate in her early twenties. The experience left her disillusioned by the "the fakeness of politics" and how personalities got in the way of solving problems for people, so she withdrew from politics for some years. Kathryn found herself engaged in electoral politics after Trump's victory, but this time she supported the centrist Max Rose.

> AUTHOR: How would you describe the situation in the Democratic Party in Brooklyn since 2016?
> KRASE: In my experience, the Democratic Party, the capital D in Brooklyn, is not visible. Since 2016, the notion of a Democratic

voice in Brooklyn is a really hard one because there's dueling parts to the Big D. In my circles, it's the progressives that have really organized around Women's March and the 2018 Blue Wave stuff. All of that is happening outside of Big D politics. It's happening in smaller progressive circles, where, honestly, they don't talk about the Big D party positively at all. They're not supported by the party at large because the party doesn't have the control over those groups. Previous to 2016, I did health-care advocacy and protest stuff, with threats to Obamacare and those kinds of things. The activism around things like health care, same-sex marriage, big issues like that, I didn't really see it coming from the big Democrats. It was coming from the smaller groups and then directly in response to Trump's election.[29]

Kathryn is frustrated that progressives—including the Working Families Party, of which she is a member—chose to run a candidate against Rose, fearing that it would spoil the race. She also feels that her progressive friends did not recognize that the new district including Staten Island has a large number of conservative voters, thus making it more difficult for a progressive candidate to win. Kathryn's fears were confirmed. The election was won by Nichole Malliotakis, a Republican who voted for Trump in 2016.[30]

The history of the political field in this chapter provides a case for NYC-DSA and its place in the international history of democratic socialist parties. Like these parties, NYC-DSA is a political and democratic alternative to neoliberal social democratic parties and their machine politics. The chapter has also brought insights into the cultural transformation of political life that are lacking in the literature on movement parties and democratic socialism. The new NYC-DSA emerged as an outside actor in the local political field, away from other flows of power and information, away from other institutions. This sociological perspective helps explain why people embedded in the tradition of the Democratic Party and the people in the new socialist movement have different understandings of the situation. Traditional Democrats feel that the new NYC-DSA lacks grounding in Flatbush, overlooking the organization's base building for years. They also do not see that the new NYC-DSA is motivated by the bigger historical events such as neoliberal capitalism and climate change. I agree that there is an element of NYC-DSA using local elections to advance translocal agendas, and it would benefit from developing closer local ties; but I also think that the push for the translocal climate agenda is important and that it had to come from an independent actor. This is also why I do not see NYC-DSA as a tribalist formation. Although it derives from indignation about alienating forces of the political system, NYC-DSA is a strategic organization focused on influencing the system, not retreating from it or fighting it from outside. In this respect, NYC-DSA is fundamentally different from Steinbeck's farmworkers or today's climate protest movement.

6

The Alexis Campaigners in Flatbush

The 2022 David Alexis campaign for New York State Senate was a high-stakes campaign for NYC-DSA. Above all, it was a pivotal push in the organization's multiyear "Public Power" campaign for the New York State Build Public Renewables Act (BPRA). This bill would eventually make New York the first state to introduce principles of the democratic socialist Green New Deal. NYC-DSA escalated the "Public Power" campaign by electoralizing it, and Alexis was up against the most powerful politician on climate, namely, Kevin Parker, chair of the Energy Committee. Parker had been delaying the BPRA for years while receiving donations from fossil fuel companies. Alexis suffered a narrow loss, but his campaign helped build pressure to pass the bill the following year (see chapter 9). His campaign had other principled goals, including to expand NYC-DSA's base into new territory south of Prospect Park and to help diversify the organization. This was a senate district with 308,000 residents. NYC-DSA knew that it would be a tough race, but the spirits were high. The organization had built momentum for years. The six climate-themed campaigns of 2022 were a culmination point in the development of the new NYC-DSA. The Alexis campaign provides insight into a crucial moment in the organization's history and shows what everyday life in the organization is really like.[1]

NYC-DSA focuses on state government elections to concentrate its limited resources on the most powerful institution within reach. The city government has less power, and the federal government is not within reach. The legislative branch of the state government, consisting of the senate and the assembly, has the power to redistribute wealth and resources. It also has "the power of the purse," with a budget of $257 million in fiscal year 2022–2023. The legislature can raise taxes, create tenant protections, and transform the energy sector. The centrality of lawmaking in US state governance explains why NYC-

DSA's climate group has a strong focus on the legislature and uses the congressional Green New Deal proposal as a discursive framework. However, New York State is not an easy and unproblematic alternative to national politics. It has been at an impasse for more than a decade. The legislative process has been described as "broken." The difference is that with the state government, NYC-DSA can win elections and work with its electeds to advance broader movement goals.[2]

The electoral campaign is a relevant setting for studying NYC-DSA because these campaigns define the organization. Campaign canvassing is the core collective practice, and electoral campaigns are more frequent than issue campaigns. Every active member in NYC-DSA is canvassing. New members are activated, onboarded, and educated through canvassing. Organizers are recruited from among repeat canvassers. People continue to canvass when they are leaders. DSA-backed electeds—the legislators who were recruited by DSA and whose campaigns were developed and run by DSA—canvass for insurgent DSA candidates, thus demonstrating their continued involvement in the grassroots and their connection with voters. I entered the organization like other newcomers by becoming a canvasser and found that it is a vantage point for understanding everyday life in the organization. This helped me get inside the organization and gain more trust than if I had just participated in social events and conducted formal interviews.

This chapter begins with an ethnography of the campaign office in Flatbush, the common space. It then moves to the level of the individual, exploring the roles of campaigners in their organizational context by providing a few examples. This investigation into the concrete acts, experiences, and circumstances of participants in the campaign is essential. It gives us a more credible understanding of who the participants are, what they do, and how they are attaching meaning to their actions and political identities. Without investigating this, we are left with vague and unfounded generalizations. By studying how people enter NYC-DSA, what they do, how they feel, and how they find common ground, we can also lay the foundation for the broader conception of the organization provided in chapter 8. My analysis shows that participants have different backgrounds and perspectives and that their political identities are dynamic.

How Did the Project Develop in the Field?

I worked inductively, approaching informants with a general question rather than with a theory to be tested. In inductive research, insights from explorations in the field lead to revision of the research question. This methodology is great for discovering new phenomena and challenging existing theories. It also means that the researcher becomes more embedded in the data. The researcher "coproduces" data with informants. The insights are gained from the researcher's reactions to situations in the field, especially from interactions with informants to understand how they perceive themselves and their social world. If another researcher had participated in the Alexis campaign, they would have responded differently and produced other findings. It is therefore important to indicate how I produced this evidence. What places did I enter, with what questions, and what relationships did I develop?

I approached my informants with the general question of how they experienced community in DSA. Sociologists have, for decades, reported on a general decline in traditional forms of community, from the nuclear family to political parties. How could the growth of new social movements in the 2010s be understood in this context? Did NYC-DSA represent an exemplary case? The pandemic scuttled my plans for a media research project (see the introduction), and I initially tried to adapt this project to DSA. The strong focus on media did not make sense to my informants. It would have been ignorant to continue unchanged. As a qualitative researcher, I am guided by the concept of *cognitive empathy*. This refers to the process whereby the field researcher comes closer to knowing how informants understand themselves. The goal is to understand more about what informants perceive, what it means to them, and what motivates them—from their perspective. Ultimately, the general question became, "What is distinctive about DSA's movement culture and its potential for influencing climate legislation in New York State?"[3]

To make the most of my fieldwork I focused on one campaign and a small number of people in this campaign and developed my ethnography of the organization from there.

Into the Campaign Office

Sunday, March 13, 2022. I have a coffee in the morning in my apartment on Tenth Street in Park Slope. My desk and computer are filled with documents about the socialist movement in the form of books, notes, *Jacobin* articles, and emails from NYC-DSA lists to which I have subscribed. Emanating from these documents are the names "Bernie" and "AOC" and terms such as "power" and "solidarity," "fighting" and "winning." I recognize the red visual identity of *Jacobin* and NYC-DSA. The websites of the electoral campaigns launched this month present culturally diverse young candidates, radiant colors, demands for justice, and the word "ecosocialism." The movement energy and positivity appeal to me in this gloomy pandemic time. I am about to start as a DSA petitioner in just a few hours. However, I fear what is expected of me because I am not sure I agree with all of the organization's political goals. They feel a bit hyperbolic to me, and I do not normally think of myself as a socialist and call for the end of capitalism. Somewhere online I read a statement about revolution. Is that NYC-DSA's goal? Can I participate authentically without compromising my values? I have other challenges: I have not participated in US electoral work before. I am not a native English speaker and do not have local knowledge of Flatbush. I am sure to embarrass myself, but that is the ethnographer's predicament—fumble, fail, and learn.[4]

I arrive at the campaign office, a one-story building in the rather nondescript area of 1179 Flatbush Avenue between Prospect Park South and Little Haiti. The northern part of Flatbush Avenue is a vibrant area of shops and restaurants, including corporate chain stores and gourmet coffee shops. But this southern part of Flatbush Avenue is not much of a destination. It is a place where residents go for some mundane services. Across the street is a grocery that has been locally owned for decades. The campaign office is located on a strip along with equally unprofitable small-scale cultural products, such as marijuana equipment, music lessons, a Haitian beauty salon, and now ecosocialism. The choice of office location is a practical decision. It is cheap and centrally located in Senate District 21. The crowd at the office did not hang out in the area. After canvasses, we would usually congregate in the backyard for a vegetarian barbecue if the weather allowed or walk a few blocks north to the FIB

Tattoo Bar and Grill. The bar was not posh, but it was a type of establishment more common in gentrified areas of Brooklyn.

After chaining my bike to a bike rack, I look around. An elderly man is walking his dog, and there is not much street life. The security grilles on most shops are closed. I open the glass door to a mostly empty campaign office. "Hi, welcome. Come on inside! We're starting in ten minutes. I'm Cam [Delaney]. I'm one of the field leads." "Hi, I'm Fabian, nice meeting you. I'm new here." "Don't worry, we'll send you out with a buddy." Cam is busy arranging printed forms for signatures. They and co-field lead Robert Wood prepare the canvassing routes for this shift. Cam is in their early thirties, energetic, positive, and focused. They have a crew cut with a straight fringe and big rimless glasses and are wearing an unassuming white sweatshirt, black pants, and black boots.

I see two other volunteers in their twenties unzipping their winter coats. They seem to be friends. I scan the room. "Yeah, we just moved in two weeks ago," says Nadia Tykulsker, who is deputy campaign manager. The walls are blank, and the only furniture in the room is an old desk and three folding chairs. The room is designed as a retail space, but it is now the common space of this campaign. This is where teams of volunteers would meet before and after their shifts, knocking on a total of 130,000 doors from March through August.[5] This is where social events such as game nights with Alexis would take place and milestones would be celebrated. There is a small back office where staff worked when they needed quiet, along with a bathroom and a kitchen. But on this day in March, the campaign was just getting started in Flatbush. Campaign manager Devon McManus took up his position two months ago.

"Hi Fabian. How are you?" says Robert, whom I met online a few days ago in an Ecosocialist Working Group meeting. He participated in environmental protests for years before joining DSA." Robert is in his late forties, and his style is similar to Cam's—a post-bohemian-graduate-student-on-a-budget style. He has very short hair, a dark-green cap, an untrimmed beard, a black puffer jacket, gray denim, and white sneakers. "It's funny that we have both been involved in music research for a long time. Are you working on climate now?" I reply that I find it hard to justify spending all my time writing about music when there is a climate crisis. "Oh, I feel the same way. I think we need people to still be writing articles on Mozart, but personally my brain just isn't there."

The room is filling up and buzzing with conversations. I recognize Will Rudebusch, whom I met last week petitioning a few blocks from Whitney Witthaus. He is in his midtwenties, a software engineer. He looks like he just rolled out of bed, with a sleepy cool appearance and short uncombed hair. He wears black rectangular glasses, a black North Face winter jacket, and a black shirt and jeans. I ask him if he has considered joining the Democratic clubs in the neighborhood. "I kind of know about those clubs, but I just want to get far enough away from that established party as possible. I just think I am inspired by other people in DSA." "Are they your friends?" I ask.

> Yes. And the existence of that entire organization is inspiring to me. There's people that work harder than me that are unpaid and are out there three times a week. It's mind-blowing because they are so enthralled with the vision and the goals of the organization. Those are the ones that inspire me. But honestly, that's because they have the time to do these things. Unfortunately, to be politically active, you can't work two jobs or be a gig worker. This is very apparent sometimes who you're interacting with. It's always people with master's degrees, which is a little embarrassing because labor stuff used to be working class. Coal miner strikes. That was people power. In American culture, college is somehow associated with political involvement now. The system is working for them.

"Hey, I'm Josh [Kraushaar]," says a short-haired guy who is passing by and joins the conversation. He wears a black jacket and also looks like he just rolled out of bed. "I'm Fabian. Where do you live?" "I live in Queens." "That's a bit of a hike. We were just talking about friends. Do you have friends in DSA?" Josh replies, "Yeah, I have a lot of people who are my friends, which I really like. This is the first time I felt like a coherent political space since college." In an interview one week later, I asked Josh how he become so attached to NYC-DSA. "I just truly believe in DSA's mission. And I think that in New York especially, there's this sense that there's a lot of stuff that's winnable here. I don't know if I was in a different chapter, how I'd feel, but I'm in the most successful and strongest chapter. I can literally see concrete victories and learn a lot more about what building power really means."

Robert calls out to the group, and everyone turns around and looks at him. People remain where they are, spread across the room, and Robert addresses the group slightly on the right-hand side. He speaks like a friendly teacher or coach. He referred to the petitioners in this room as "Team Sunday" in an email a few days prior. National DSA routinely addressed members as "comrades," evoking the sense of a political public, but the campaign referred to us as "teams," as groups tasked with work.

The opening ritual at the beginning of each shift is designed to welcome, inform, and coordinate. The atmosphere is focused on the task ahead but relaxed enough that people are comfortable asking questions and making brief comments. The orientation by the field lead is followed by a roll call, in which each canvasser acknowledges their presence and is acknowledged as a team member, with the group nodding or cheering. The ritual is somewhat improvised. The name of the organization is rarely mentioned, and there is no display of its visual identity, except that a few participants wear a DSA T-shirt (campaign T-shirts became more common at the end of the campaign).

The social media communications of the Alexis campaign mostly reached existing NYC-DSA members, just as most of the donations came from these members. The campaign's 287 posts to Instagram, the main social media platform for the campaign, included many photos of canvassing teams to boost participation and movement identity. Most of the posts had explicit political messaging, frequently with an agitative and combative element, often including the word "fight." A few posts highlighted housing, health care, and climate, but climate was generally not front and center of the posts. The movement approach becomes clear if we compare this campaign with the Kevin Parker campaign. The Parker campaign did not have the same large canvassing operation and posted less on social media. It was a somewhat passive campaign. Thirty-nine posts were made to Instagram. One of them was a photo of canvassers. The other posts were corporate-style photos of Parker in a dark-blue suit and videos announcing the election date, along with claims about his achievements, to a soundtrack of smooth jazz. The campaign had a larger advertising budget, so the content reached a larger audience; and the campaign sent out more mailings.[6]

On this day in March, Robert begins the welcoming ritual:

Hello everyone! Thanks for coming out today. I'm Robert. I'm a field lead with my colleague Cam. We're super excited about this campaign. We need twelve hundred signatures to get David on the ballot, but our opponent is going to challenge our signatures. So we're aiming for three thousand signatures, and it is important that the signatures meet the board's requirements. Make sure that the full name and the address are written in capital letters in the first field and the signature and date right below. We'll do a short training session for those who haven't petitioned before, but basically you just have to ask the person if they are a registered Democratic and want to help get David on the ballot. Let's start with a roll call. Please say your name, where you live, and why you're here.

The team responds: "I'm Josh [Kraushaar]. I'm from Queens. I want an ecosocialist in office!" A few people nod in approval. "Nell [Crumbley], Flatbush. Yeah, I'm excited about David's campaign!" The group looks pleased. "Yeah, I'm Daniel [Goulden], also Flatbush." "I'm Stephanie [Lemieux], Flatbush." "Ben [Lenz], Flatbush." "Hey, I'm Andrew [Butler]. [dramatic voice] Fuck capitalism!" The group responds with approval and laughter, and one person loudly exclaims, "Hell yeah!" "I'm Fabian, Park Slope, um eh . . . I'm here to support David." "Theresa [Paquette], Crown Heights." "I'm Will [Rudebusch], Flatbush, and I think David is a great candidate." The group cheers. "I'm Sarah [Reibstein], Crown Heights." There are a dozen volunteers today, plus the field leads and campaign managers. The group is large enough to create a collective identity yet small enough to provide a personal and informal setting where it is easy and natural to interact with everyone.

I am not a citizen and therefore not allowed to petition alone, so I am paired with Josh. We are assigned a stretch on Cortelyou Road. On our way to Cortelyou Road, Josh and I quickly get into an exciting conversation. I am eagerly asking questions, and his answers are insightful. We have mutual interests on a deep level, and Josh is teaching me new perspectives on politics that I find meaningful. Within a few weeks, I would begin looking forward to walking with Josh and Will every Sunday and talking to them about DSA and politics. I was not the only campaigner who valued such conversations. I also enjoyed the direct and playful style, the irony and sarcasm about neoliberalism, that would not have been appreciated in our respective workplaces. Campaigners took inspiration from the socialist podcasts mentioned earlier (see chapter 4), but the culture among campaigners was a culture of embodied collective action structured by the material environment of Flatbush. The campaign offered a refuge from the alienating forces of corporate capitalism, the mass-media world of politics, and, for me, neoliberal higher education. I felt drawn to the cultural resources of this movement space and gradually understood more about the movement culture. In the process, I became politicized and attached to NYC-DSA.

For a couple of hours, Josh and I interact with hundreds of pedestrians on Cortelyou Road. "Are you a registered Democrat?" and "Hi, do you know there's an upcoming election?" are some of my opening lines. Standing in the same place for hours and trying to reach as many people as possible, I am exposed to the area's diverse population and people's attitudes to politics. Many have no interest in politics, and the conversations are brief. So petitioning is boring and tedious at times, but it is necessary.

I felt uplifted when a stranger recognized the civil society perspective in this work, and this happened more often in canvassing, where the conversations were longer and deeper. Over the course of my canvassing experience, a few constituents would say something like, "I see what you're doing. This is valuable for civil society. Thank you." This feedback, even if rare, meant a lot to me. Canvassing provides plenty of opportunity for the individual canvasser to develop meaningful conversations, and I agree with my colleagues in NYC-DSA when they say that canvassing helped them restore faith in civil society.

After two hours, Josh and I count the signatures and continue the conversation at the campaign office and the postshift barbeque. Nadia takes a photo of the team and shares it with the campaign's volunteers the next day via email. This group-photo ritual enhances the sense of community and belonging to the campaign.

The weekly team photos tell the story of the campaign's trajectory. The first photos of team Sunday feature a small group in cold March. The teams grew larger when the campaign moved from petitioning into canvassing, the weather got nicer, and Election Day got closer. Photos from July and August show crowds of forty and fifty canvassers (figure 6.1). More people from the neighborhood got involved, including more

Figure 6.1. Canvass team at the David Alexis campaign at the Brooklyn Public Library in Flatbush, August 2022. The photo documents the typical integration of participants with different roles, including leaders, cadre members, casual canvassers, and electeds. There were not always state electeds, as is the case here. From left to right, Nadia Tykulsker (kneeling front row), Aina Lakha and Sarah Reibstein (slightly to the right of Tykulsker, third row), David Alexis (seated front row), Will Rudebusch (standing against the wall), and Devon McManus (standing, with both arms in the air holding fliers). (Photo from the campaign's Instagram account)

people of color. There was a group from New York Communities for Change that canvassed on their own, but that was more because of the self-determination of their local leader, Leroy Johnson, and not because of racial boundaries around the Alexis campaign. Most DSA members in Flatbush are white transplants and not from the Caribbean diaspora, but the campaign teams eventually became racially diverse.[7]

From my canvassing and street ethnography in Flatbush, I learned that beneath the heavy fog of political apathy, many voters have grievances about matters of public concern. They just did not always develop the interpretation of their grievances into political claims, and they did not feel that they could have meaningful influence on the political process. For the same reason, constituents often did not take much interest in the list of issues on the campaign's fliers. This is a reminder that a campaign platform is an aggregate, a structured collection of issues attempting to reach a construed demographic around generic concepts of parties and programs. Not a single voter looked at the campaign flier for the first time and said, "Sure, yes, these are my exact issues and priorities."[8]

Roles and Perspectives

What Do the Canvassers Have in Common?

The people in the Alexis campaign organization have diverse backgrounds, but there are also some structural commonalities. Most of the campaigners in my Sunday teams are in the same age group (twenty-five to thirty-five), and about half are not in long-term relationships. The regular canvassers generally do not have children. Many are graduate students or public-sector service workers (e.g., healthcare workers, schoolteachers, and social workers) or project managers in small independent arts organizations. There are also artists and information technology workers (e.g., software engineers and data analysts). Many have humanities and social science degrees from Ivy League schools, mostly from areas such as the arts and sociology and not administration or business. The campaigners do not belong to the managerial class of people who are hiring and managing workers. They can be identified as an urban subelite—a group of people who are close to the elites in the city, educationally and geographically, with elite resources such as cultural capital but without elite jobs and wealth.[9] They are intellectually and morally engaged, and their humanistic training also registers in critical thinking about capitalism and institutions of formal power. The campaigners channel their resources and the energy of their urban environment into the movement. New York City is a dynamic environment, with a high concentration of talent in a small geographic area. New York and Central Brooklyn in particular also have a concentration of young people with culture-driven life choices and flexible family and work arrangements, who are therefore inclined to volunteer for a movement. Also, for people with their educational backgrounds, a high-paid job in politics is not an option.

Their cultural capital is a great resource for NYC-DSA. It helps explain why the organization has developed a holistic approach to political organizing, fostered meaningful participation, and created a vibrant visual identity.[10]

How Do the Canvassers Differ?

On the basis of my observations in the campaign and in meetings of the Central Brooklyn branch, photos of canvassing teams, and the composition of working groups, NYC-DSA in the early 2020s is gender balanced and culturally diverse. There were more whites than people of color overall, but with variations. In the Alexis campaign, for instance, the canvass teams were predominantly white at first but became increasingly diverse. With some shifts in the final months, about one-third were people of color. As for NYC-DSA as a whole, Aina Lakha, one of the most active members from 2019 through 2023, described how the organization had become more diverse. Lakha remembers their first large meeting of the Central Brooklyn branch in January 2019: "I had never been around so many people who were white young professionals, young people in their midtwenties to early thirties who were white-collar professionals. It was very striking. There was definitely diversity, but for my experience in New York, it was a little different in that way. There are a lot more leaders of color now." More than half of the four thousand who joined NYC-DSA in the months after October 7, 2023, are people of color.[11]

I generally developed closer relations with people who were white and male like me. Over time, I recognized this bias and took steps to rectify it. I also became more attentive to the quiet and shy canvassers and those who had mixed feelings about canvassing or DSA and in some cases quit for that reason.

In the egalitarian movement spirit, leaders refrained from introducing themselves as such during canvasses. This led to funny situations early in my fieldwork. In one conversation, I chatted up a person who was standing next to a table with some fliers, checking her phone and not looking busy. She embodied the joyful and slightly hipster-oriented urban attitude among organizers in the organization. She was dressed casually with a black hoodie and a baseball cap saying, "I am culture."

We chatted for two minutes before one of my questions prompted her to tell me that she was a member of the New York State Assembly.[12] Leaders and elected officials were ascribed some status and felt some sense of entitlement; but the culture was quite egalitarian, and no one talked down to others or framed them as subordinates.

The following individual-level exploration helps nuance perspectives among the campaigners. I focus on their organizational roles and identities in the Alexis campaign.

Campaign Managers

Devon and Nadia were the only ones working full-time on a salary. The remaining staff members were volunteers: a dozen field leads running petitions and canvasses (two field leads per shift), a social media manager, and a graphic designer. The campaign managers were not directly hired by or paid by NYC-DSA, but the campaign was funded by the citywide campaign committee DSA for the Many and direct donations.[13] Devon and Nadia had been active in the Ecosocialist Working Group (ESWG) for years and played a key role in electoralizing its "Public Power" campaign. Devon was also involved in DSA for the Many, which supported the proposal for running six ecosocialist campaigns in 2022. The Alexis campaign was formally developed by the Central Brooklyn Organizing Committee and Electoral Working Group. It had been approved by branch members in an online vote. Informally, however, the campaign evolved from the "Public Power" campaign, which began in 2019. The ESWG had developed organizational capacity and essentially took over NYC-DSA's electoral operations in 2022.

Daniel Goulden has organized with ESWG in Flatbush since 2019 and gives a sense of the emerging campaign organization:

> We love our socialists in office, but this time we would like eco people to be part of the race, someone who's super knowledgeable of climate issues. Everyone loved David. Everyone wanted to challenge Kevin Parker. Grace [Mausser of the Central Brooklyn Organizing Committee] and I were the ones creating and leading the kitchen cabinet for David's campaign. In NYC-DSA, a kitchen cabinet is a kind of protocampaign of eight to

fifteen cadre members. We prepared his endorsement process, created a website and a social media presence, and boring shit like that. I recruited Nadia to be on the kitchen cabinet. That's how she got to know David and got involved and ultimately became deputy campaign manager. The kitchen cabinet essentially dissolves when you hire campaign staff. We hired Devon in January, and then I was elected as cochair of the Electoral Working Group.[14]

Devon and Nadia assumed a backstage role in the Alexis campaign. The campaign's voices in the office were the volunteer field leads. Canvassers regularly communicated in their own words to subscribers of the campaign's email list. The managers instructed field leads and created infrastructure. They created a sense that NYC-DSA takes electoral campaigns seriously and does not want to turn them into a spectacle. Devon and Nadia instilled respect for the democratic process and the rules and did not demonize the opponents.

Nadia Tykulsker is in her early thirties, single, and from Brooklyn. She graduated in dance at the University of Michigan and moved back to Brooklyn after teaching for a few years. She has worked as a dancer and choreographer on the Williamsburg scene and as an arts administrator for the nonprofit FABnyc on the Lower East Side. She has been involved with DSA since 2018 and the ESWG in particular. She was one of the ESWG organizers who participated in a protest that escalated the "Public Power" campaign in 2021. Together with ESWG organizers, she published an article in January 2022 questioning Kevin Parker's moral integrity: "State Senate Energy Committee Chair Kevin Parker receives the most fossil fuel money of anyone in the legislature and uses his position to block every climate bill. Who needs Republicans when you have the Joe Manchin of New York?"[15] Nadia has a big, friendly smile and is always polite, but she is very work focused and a little reserved until she recognizes a serious long-term commitment. For these reasons, I connected with her at a late stage and did not manage to recruit her for a formal interview. In the Alexis campaign, Nadia worked closely with Devon, but she did more of the budget and digital communications, while he did more of the strategic overall management.

Devon McManus is in his late thirties, has a girlfriend, and has been living in Brooklyn since 2009. He had just moved to Flatbush after becoming manager of this campaign. Devon graduated from college in New Hampshire in 2009 and moved to Williamsburg to be part of its indie music scene. He worked for touring bands before becoming a sound engineer at the Union Pool club. He also did sound at a few experimental grassroots venues where I conducted research from 2010 to 2013 for my book on the music field. Devon left the music business in frustration: "I think this work is really hard to do well without being an asshole. I was getting like 10 percent of what the bands made, which was basically nothing." His path from music to climate politics is typical. Many participants in Williamsburg's independent arts scene became engaged in NYC-DSA. The scene was decimated by gentrification, and participants got older and responded to national political events. Devon became a Bernie fan. He worked for the international advertising agency Blue State for years, but he was not passionate about it. He refers to the job as a nice financial cushion, and it allowed him to work on DSA stuff for hours every day. Devon got into the new NYC-DSA and its electoral work right from the start in 2017 and became friends with Tascha Van Auken. The two managed Julia Salazar's 2018 campaign. Devon is one of the most experienced campaign managers in NYC-DSA and has held various leadership positions. He is also the main architect of DSA for the Many.[16]

How does Devon compare the Salazar and Alexis campaigns?

> Julia ran against Martin Dilan, who was also a pretty good villain [referring to Kevin Parker]. But the races are so different. Julia's campaign happened when Trump was in office and everything was super awful, and AOC's victory made people want to be part of the next socialist victory, and Julia was a Latina woman, like AOC. The neighborhoods are different. Julia's was in Bushwick, Williamsburg, Greenpoint. David's is in Flatbush, where the vast majority is Black and there's not a lot of millennials. But it's also a different candidate. David didn't go to Columbia like Julia did. We had to work with him a lot on public speaking, but it's really hard when you have a full-time job on top of being a candidate.[17]

Devon got politically activated by the 2016 Sanders campaign and is in the faction of NYC-DSA closest to Sanders, the Socialist Majority. He is a humble and quiet person. He is reflexive but has little interest in the intellectual side of politics. This became clear when I asked him about his media repertoire:

> McManus: I'm not a big reader. I read a bit before bed or in the morning. I listen to a lot of the BBC just because it's on the radio. And it's nice to just hear someone British, even if they also have frustrating politics at times. My family and older friends watch a little too much MSNBC, which has turned into like a soap opera at times. Obviously, I'm not a big fan of Fox News. *The New York Times* and *The Washington Post* can be a little frustrating.
>
> Author: Does one have to check Twitter to be informed about NYC-DSA?
>
> McManus: No, definitely not to be informed. I think it's a bubble. In New York, politicians spend an insane amount of time on Twitter. In a weird way, it's important for Albany [state government] stuff. I think there are people who get their news that way, so it's helpful for reaching them. But that's not where their organization needs to grow.

Many but not all in NYC-DSA are antisectarian and antiableist, and Devon's perspective brings nuance to this perspective. His relationship with the term "socialism" is shaped by his goal of helping build a broader Left that includes unions.

> Author: How do you translate ecosocialism into this campaign in Flatbush?
>
> McManus: A good example of that is that a lot of DSA members are supportive of David because Kevin Parker is the energy chair and climate change is a big issue. On the doors here in Flatbush, I've had conversations where people say, "It's not my thing. What I really care about is my rent and my health care." Those issues make a huge positive impact on people's lives, but they too are part of our platform. The more immediate concerns tend to be easier to address. This is something the climate movement struggles with in general. Once people cross the threshold and are, like, "Oh my God, this is the thing that will end the world, and we'll do everything possible to stop it." It takes a while to get there. But it's a pretty big jump to

go from being fired up about this issue to wanting to join a socialist organization, especially one that in my opinion has a bit too much of a fixation on the word "socialist."

AUTHOR: And socialism has different meanings to different demographics, right?

McMANUS: Totally, and I think it's important for us in DSA to be humble and not like, "It's, like, our way or the highway." We need to work in coalition with these other groups.

AUTHOR: Is the term "socialism" a barrier to such coalition work?

McMANUS: From my perspective, it's like, "Who cares about the label?" Just to be realistic, DSA is in the place it is today because Bernie Sanders was a democratic socialist. It's not like the label doesn't mean anything, but a few years from now, hopefully, there'll be another Bernie figure. We shouldn't get like, "Oh, we're not going to support him because they won't take on this moniker or something." As long as they're, like, working towards justice and equality in a way that's sustainable and a nonreformist reformist, like you might be doing incrementalism, but you're doing it with an eye to—you're not settling for this, and you're doing it in a way that makes it easier for working-class people to get involved in political office. It creates these coalitions of people with interests, like tenants are a great example or people who use health care, like these broad swaths of the population. And there's not just DSA. There's also the labor movement, which is a huge political force that isn't aligned with us a lot of the time.[18]

Field Leads

Field leads organized petitions and canvasses. They planned the routes and communicated with volunteers in the office and through the weekly postshift email with a team photo and call to participation. Being a field lead involves a considerable amount of work and is not something that newcomers could do. Halfway into the campaign, one of the field leads was burned out. The role requires a lot of passion because it involves many hours of unpaid work with little status attached to it. However, field leads are rewarded by the sociability of volunteers and by getting closer to the organization and the candidates. In the following, I explore the perspective of the field lead I worked with the most.

Robert Wood is in his late forties, married, and a freelance political writer. He studied piano and has taken an indeterminate break from his doctoral studies in musicology. While in graduate school in the 1990s, he read much of the work of Karl Marx and Theodor Adorno. Robert began working as a marketing copywriter for music nonprofits, writing program notes for ensembles, and eventually transitioned into freelance journalism, specializing in politics.

Robert first became politically active through environmentalism. He participated in pipeline fights throughout the 2010s, nationally and in New York City with Bill McKibben's 350.org. The Keystone Pipeline protests in Washington, DC, in 2011 were a formative experience. He met James Hansen and Naomi Klein there.

> For some reason, the pipeline fights always got my attention. They're very tangible examples of how capitalism infringes on people's lives. Pipelines directly destroy communities, environments, and ecosystems, whether it's into the water they're passing through or native land or whatever. There's something satisfying about a tangible object you can fight. I viscerally hated the idea of the Keystone so much. The Dakota Access Pipeline situation was happening around the same time, and I started reading a lot about Native history. I read Dan Brown's *Bury My Heart at Wounded Knee*, a classic history of the West told from the perspective of the different tribes. That really activated me towards the Dakota Access Pipeline project, but I never went out there, and I regret that.[19]

Robert moved from this perspective of local community protest against capitalism to DSA's movement party project. He moved from 350.org to DSA to be part of the "Public Power" campaign. DSA's political framing of climate resonated with him, and he had several encouraging personal encounters with DSA members in Brooklyn. Robert's continued focus on environmental and climate issues is reflected in his media diet, with a particular interest in publications such as *Grist*, *Inside Climate News*, and *The Smog*.

Casual Volunteers

Organizers in NYC-DSA tend to cultivate the ideal of the passionate volunteer to boost participation and develop participants into leaders. However, many campaigners were casual volunteers. A few dropped out after a few canvasses because they were not comfortable talking to strangers about politics, and one canvasser withdrew after a racist assault.

I define the casual canvasser as a role that does not require involvement beyond the few hours of a weekly shift but still creates a meaningful experience of political participation. It creates the experience of contributing and talking to constituents. Some casual canvassers approached the weekly shift as an alternative to spending the Sunday afternoon at the TV set or going to a café or museum. To them, this was a more meaningful way of spending Sunday afternoon.

Sarah Reibstein is in her early thirties, recently married, and a doctoral student in sociology at Princeton. She grew up in Westchester County, just north of New York City, to parents working in the city's journalism world. Her father has been an editor or writer at *The Wall Street Journal*, *Newsweek*, *Forbes*, and *Bloomberg*. Both parents are Democrats but not politically active. Sarah remembers watching *The Daily Show* with her father in high school and hearing him talk about George W. Bush. She became more conscious of imperialism as a visiting college student in Ghana. However, her political engagement did not fully develop until after she graduated.

> In my economics classes at Northwestern, it felt like it was all about maximizing profit, and that didn't seem good. I thought that it could all be solved with behavioral economics or some other more humanistic gloss on capitalism. It wasn't until after college that I came to the realization that capitalism is really the problem. I started going to bookstores and furiously Googling the internet and did a lot of reading on my own. Me and my boyfriend [a Columbia graduate student] became involved in the New Economy Coalition, not socialist yet but anticapitalist. I would go to Richard Wolff's talks at NYU and later joined the Post Growth Institute. I began researching universal basic income.

Sarah's primary political engagement is with the graduate student union at Princeton. She feels that just talking about capitalism made her feel disempowered but that "the union project was very concrete and right around [her]." Her interest was triggered by a town hall where a talented organizer related moving testimony about her problems with the health-care system. "I was like, 'Oh my God, Princeton is not actually great for everyone!'"

Sarah joined DSA after moving to Brooklyn in February 2020. She was not involved in the Sanders campaign in 2016. She donated to his campaign but voted for Jill Stein of the Green Party. She also did not feel connected with the anti-Trump movement, even though she was deeply affected by Trump's election. "It was awful. I was trembling," she recalls. In retrospect, Sarah finds it odd that she did not get into DSA earlier but says that living in Princeton made her somewhat remote from national politics and DSA. The DSA movement did not have a strong presence there, nothing like in Brooklyn. The "hashtag activism" of the anti-Trump movement felt "shallow" to her. Her thoughts on her level of involvement with DSA are insightful:

> AUTHOR: Have you considered joining some of the working groups in DSA?
>
> REIBSTEIN: I think the working group stuff is important, but I decided I don't have time for it because the union organizing is my main thing. Canvassing for David's campaign just feels like something that's manageable. I can plug in for a few months and feel like making a difference. The last one in 2021 was Michael Hollingsworth for City Council, which we lost. So coming to terms with spending all that time on a thing that then you lose, and then it doesn't mean anything. . . . I thought David has a good enough chance that it was worth trying again.

Casual canvassers such as Sarah spoke more freely about DSA than did the leaders, who were more invested in the organization's identity.

> AUTHOR: How would you describe your attachment to DSA? And do you feel that you are blending in with the crowd?

REIBSTEIN: I don't feel like a real insider because I feel like I haven't earned that by being involved. But I do think culturally, I pretty much fit right in a way that's kind of embarrassing, you know? I'm like the target DSA. I live in Brooklyn. . . . I feel like everyone that I meet there, it's like we've dressed the same, and I'm like, yeah, these are my people lined up. It's a little embarrassing because no one wants to be a type or anything. But on the other hand, we get blamed for being gentrifiers, but we're doing all we can to stop the neighborhoods from gentrifying. I wish the organization represented a more diverse working-class population.[20]

With regard to Sarah's news media diet, she listens to *Democracy Now!* in the morning and feels that it gives her a perspective that she has been missing in *The New York Times*. It gives her a sense that the Left matters. She occasionally reads *Jacobin* and likes some of the recent political writing in *Teen Vogue*. She has mixed feelings about *New York* magazine but canceled her subscription when she was politicized. She describes *The Atlantic* as a centrist publication that her parents are reading.

Like everyone else in this campaign organization, Sarah contrasts with the movement warrior type celebrated in pop-culture images of social movements. She talked little and with a soft voice and was not focused on labeling or slogans. She did not put herself in the center of attention.

When I reconnected with Sarah in 2024, she had finished grad school and had gotten much more involved with DSA. This process began in the Alexis campaign, in which she would eventually lead phone banks and transcend the role of a casual canvasser. Two years later, Sarah was a member of the organizational committee of the Labor Working Group. This trajectory illustrates how the organizer development model of NYC-DSA helps casual canvassers move into different categories over time.

William Rudebusch is in his late twenties, recently married, and a software engineer. He grew up in the country, studied mathematics at a college in Sioux City, Iowa, and taught there for a few years before moving to New York out of personal interest. He appreciates the critical mass in NYC-DSA. Will grew up with parents who were not academics and listened to Rush Limbaugh, one of the most popular neoconservative radio hosts in the 1980s and 1990s. Limbaugh declared himself "liberated from the East Germany of liberal media domination" and railed against feminism, environmentalism, and climate science.[21] Will was first politicized by the 2020 Sanders campaign, when "everything came collapsing and people were losing their mind that Trump was president and calling women fat and stuff." The encounter with the now larger movement of young socialists around AOC "sealed it" for him. At the time, he lived in Crown Heights, where DSA has a strong presence. He went to a couple of meetings and joined the support campaign for Sanders and the Phara Souffrant Forrest campaign. Will is clear that his involvement is limited. He attempted to attend weekly meetings in DSA, but he feels he lacks the time. He works full-time and is renovating the house that he bought in Flatbush last year. "I don't know how they do it. The people that run this thing must have something other than a nine-to-five job."

Will has a friendly and unassuming personality. He does not become agitated, even though he is indignant about the country's political situation. He thinks that inequality has reached grotesque levels and talks sarcastically about establishment Democrats. Will is primarily interested in national politics. Also, his political interest does not derive from personal struggles. He has a well-paying job, owns a house, and does not voice personal grievances.

> The biggest draw for me was Bernie Sanders—his demands for free health care, a higher minimum wage, and other stuff that will help everyone. Whereas if you listen to Obama, Hillary, or Biden, it's just like, "America is about opportunity." ... No one talks about NATO and its purpose. Another good litmus test is charter schools. It's a neoliberal hell! Then I can see why people check out. Yeah, "We're gonna cut funding to your school and drill for oil in the Arctic." This past week Biden just said, "If you are worried about gas prices, go buy an electric car." A $16,000 electric car? What planet are you on? The only growing demographic in the Democratic Party is affluent college-educated women. I really think about this quote by AOC where she says, "In any other country, I would not be in the same party as Pelosi."

Will's political perspective is made clearer by his feelings toward news media. He does not go to the landing pages of news media. Instead, he reads news posted by his contacts on social media platforms. He likes reading the critical responses from Reddit users. Will is inspired by the young, male, leftist commentators for independent podcasts such as *Chapo Trap House*. Those commentators excel in indignation and sarcasm, slamming opponents, and challenging the credibility of news media. "I would never before have thought about the credibility of the people who write editorials for *The [New York] Times*, like Bret Stephens. Complete hack! Loser. He was hired as the one climate denier at the paper's editorial staff. Why? Do we need a climate denier? Okay, if you're gonna hire a diverse staff, get a communist or socialist [laughs]. He says what the owning class of *The Times* wants him to say. They platformed him, and now they're publishing all this, like, 'manufactured consent' sort of stuff."[22]

Pundits

Some canvassers operate as a kind of socialist Twitter/X pundit or influencer outside canvassing. They are into independent leftist media and share indignation with people like Will, but they are more actively debating with fellow canvassers and participating in social media debates in the DSA Twitter/X public. They are casual canvassers but are more engaged in reading and discussion. They tend to have academic expertise in politics. These "canvassers with expertise" speak with more confidence about political history and philosophy and are more informed about the history of NYC-DSA. But they, too, emphasize practical organizing and canvassing and do not think that members should just spend much time on political education. Many of them have served on a working group organizing committee. Examples of "the canvasser-pundit" hybrid include Michael Pollak. Another hybrid is "the organizer-pundit," examples of which include Aaron Eisenberg and Josh Kraushaar. The public demonstration of good political judgment and organizational expertise is a defining feature of the people with top-tier leadership responsibilities in NYC-DSA, whom I address in chapter 7. Citywide leaders such as Aina Lakha, Cea Weaver, Eric Thor, Jack Gross, Neal Meyer, Sam Lewis, and Susan Kang have written articles that shaped the organization's outlook and direction, its internal democracy, or its actions in specific fields of activity.

Michael Pollak is sixty-three and single and has lived in a cheap rent-controlled apartment on the Upper West Side near Columbia University for decades. He used to teach sociology at a college in New Jersey but decided to leave academia fifteen years ago after his department was closed. Michael became a secretary and is happy about the flexible conditions of his new job in Manhattan. He is passionate and joyful. Michael loves social life and the arts in New York City. "I don't have a TV set. New York is my television," he says. He was a leftist from his late teens but went to college in a small town that did not have many leftists. He describes the final decades of the twentieth century as an era when there was no mass Left to get involved with, mostly just sectarian groups and demonstrations. "Basically, we went to demonstration a lot." Michael spoke with Michael Harrington a few times and liked his idea of a left wing of the Democratic Party in the wake of Ronald Reagan's election. But he did not join DSA until 2017. His experience demonstrates the centrality of canvassing and how it shapes the experience of political culture.

> With Sanders, an organized Left emerged for the first time in decades, and that's when I joined DSA. DSA is the only thing between the Democratic Party and the sectarian wasteland, the only nonsectarian party. So I joined.
>
> The first time I really knocked on doors was for Julia Salazar. And that's because it was the first time in my life that I thought I would be working on a socialist campaign where someone could get elected. Although I felt really good about it, I didn't think we're gonna win. But AOC and Julia both won overwhelmingly. The next set of campaigns were in 2020, when we switched to phone banking because of the pandemic. People normally wouldn't pick up the phone, but they picked up because they were home alone, and it was great having these conversations.
>
> You see, canvassing is the only thing that restores my faith in people. I mean, on the internet, everybody's an asshole. Yeah. And then you knock on doors, you realize, look, the mass of people, they've got nothing to do with that. They're not on the internet. They don't care about it. They don't talk about politics. That's our main problem.

Michael prefers to do just a weekly canvassing shift, but he has canvassed at least four electoral campaigns and three issue campaigns, adding up to thousands of conversations at the door over the course of five years.

Michael gave detailed and insightful answers to my questions, as one might expect of a former professor. He placed NYC-DSA in local and national political history and provided insights into its strategy and finances. He explained how the millennial transformation of the Left was preceded by a transformation in national leftist media magazines in New York City. First-mover democratic socialists took over magazines, such as *The Nation* and *Dissent*. Michael had personal relationships with editors and journalists of some of these publications, including Bhaskar Sunkara.

Michael shares his expert knowledge in conversations with organizers in NYC-DSA, and his activities on Twitter/X are followed by many. But he does not engage in online debates and is more interested in sharing insights from his canvassing experience: Michael explains that canvassing is more powerful than media in state government elections because voters do not expect to hear from candidates in the media. This is different in city government elections. Political scientists have noted that state political news is mostly "not sexy," unless it is about a scandal. Pollak notes that mailings, too, have limited impact: "The mailings can do a little bit, but they are totally outweighed by someone knocking on the door. When I canvassed for Julia Salazar, so many fliers were sent out every week. It didn't do any good. Nobody reads them! If voters had any questions, it was easy to dismiss it. It was hilarious! You realize just how much more powerful the doors were."[23] Michael shares the same fundamental political view of his fellow campaigners, but unlike some of the others, he is not shocked and indignant. He has experienced a longer history of inequality and political celebrity before Trump. His conditions have been stable for decades and not affected by Trump. Michael thinks climate change is important, but he is not organizing with ESWG; and NYC-DSA's base building in Flatbush is an important reason for him to go all the way to Flatbush to canvass every week.

Working Group Organizers

It took about a month before I learned just how many of the campaigners were ESWG organizers. They included campaign managers Devon and Nadia, field director Tefa Galvis, field lead Robert Wood, and canvassers Gustavo Gordillo, Daniel Goulden, and Josh Kraushaar. Daniel, Gustavo, and Nadia also served on DSA's national Green New Deal committee. Basically, ESWG organizers in Flatbush had temporarily suspended the group's activities to form the core of the Alexis campaign. They moved from a thematic working group setting to an electoral setting to meet this temporary need. This was necessary because the organization would otherwise have been lacking experienced organizers to run the Alexis campaign. But it also allowed ESWG organizers to follow through on the work for a socialist Green New Deal in New York State that they initiated in Flatbush in 2019. ESWG organizers were also more informed about climate than were organizers in other parts of the organization. They were more invested in the campaign because they had developed it. In general, organizer-canvassers were more informed and integrated with NYC-DSA than were the other canvassers. The strengths gained from teamwork—communication, community, and trust—helped them keep their grievances about politics and internal organizational tensions at arm's length. This explains why some of them had less indignation, even though they were very concerned about climate.

Daniel Goulden is in their early thirties, recently married, lives in Flatbush, and teaches creative writing at a local college. They grew up in Brooklyn in a family that had been leftist for generations. Their grandfather was in the Communist Party and was hurt badly by the Joseph McCarthy trials in the 1950s. Daniel's mother was a civil rights lawyer and taught them about Marx in seventh grade and took them to several demonstrations against the Iraq War. Daniel soon identified as a democratic socialist. They went to college at Brandeis because of its history of social justice activism but were hugely disappointed by how this tradition was being suppressed. "Brandeis had become submerged in this kind of unfettered capitalism that dominated the country at the time," they said.

I participated in Occupy and think that it really broke people out of this idea of neoliberalism and "the end of history." But a lot of the people were just really mean, and the fundamental problem was this idea that "we can't get our hands dirty with the actual politics of building power because that's a betrayal of this movement." There was this tyranny of structurelessness. DSA is completely different in that respect. A lot of leftists moved to, like, cultural criticism, identity politics. . . . It was mostly like, "This celebrity said a problematic thing."

I was pretty disillusioned when I graduated and moved back to New York. I didn't have many ways of connecting. I marched with Black Lives Matter, but I wasn't organizing with them. When Bernie ran, I started to believe that he could get people to care about democratic socialism, and I got horribly depressed when Trump won. I organized with Indivisible, which was fairly liberal politics. Actually, there was no politics there. All they cared about was Trump.

I moved to Africa for a year, went to grad school at Cornell, and got involved in some organizing there. But I first got involved with DSA when I moved back to New York in 2018. AOC and Julia Salazar had just gotten elected, and they were doing so much cool stuff. They stopped the Amazon headquarters project. Eventually, I found myself with ecosocialists, and Gustavo Gordillo started giving me stuff to do. I organized town halls in Flatbush to target Kevin Parker, who has now been the top target of the ESWG for four years. He gets more money from fossil fuel companies than even Republicans, even though he styled himself as progressive. To be totally honest with you, I've had pretty intense climate anxiety since I was a kid. My father has always been anxious about environmental issues. I really liked everyone who I met in the Ecosocialist Working Group. Unlike my college experience, people were just nice, and they were giving these tasks to do, and it was fun.[24]

Like many other members with a history of activism predating 2016, Daniel experienced NYC-DSA as a well-ordered and positive environment of constructive political organizing. Daniel was the strategy cochair of the ESWG when Alexis was recruited and played a key role in the recruitment.

Josh Kraushaar is in his early thirties, has a girlfriend, and is a data analyst living in Astoria, Queens. He thrives in the lively conversations around the campaign office, sharing his sardonic political musings and listening to friends and strangers. Josh is an organizer-pundit hybrid. He is known as a clever and uncompromising commentator on Twitter/X (this was in 2022). He is not afraid to call out unethical behaviors among DSA members. His Twitter/X handle is aptly named "Hot Take Sommelier." Josh was already a fearless and indignant critic as a middle schooler, when he published articles about the Iraq War in the local newspaper. "I got one or two published, and then they heavily censored one that provided death counts in the Afghanistan and Iraqi US invasions. That was a total big break for me. There was a period when I didn't trust any of it."

Josh is a voracious reader of political literature, including books on the history of the welfare state, labor unions, social democracy, and movement organizing.

He grew up in a suburb of Albany to liberal parents. His mother was a social worker in the Department of Labor, his father an administrator at the State University of New York. Socially, however, the family home left him with a painful experience of isolation. "One thing that always really haunted me was how lonely my parents seemed. Even when I was in my early teens, I was like, 'Wow, they just don't see people ever.' They'd go weeks without seeing friends, even when my sister and I were old enough that they didn't have to baby us 24/7. That was rather grim." Another formative experience was studying economics in college. Josh did not know about leftist politics when he was in high school and chose economics because he thought that it was "a liberal field" and a vital current in the opposition to George W. Bush. His understanding changed: "American economics is very bad. It's like pure neoliberalism. Inherently evil. I hate to say that I studied economics. Satanic discipline! [laughs]." After graduating college in 2016, Josh started feeling lonely and "fucking miserable" because he was working all the time, and it was hard to meet new friends. This was when the Sanders/Trump moment happened, and Josh became more engaged in politics. He tried to join political groups in Boston, but they were "hyperlocal and not great. There was no broader current." This changed when he moved to New York. "I moved to the city in 2018, when AOC had just won. I was like, 'Oh, this actually has got something to it.' I started to get active that way. AOC's victory was, like, the first Left victory ever, and I think it's why DSA is so disproportionally successful in New York City—and the history of the Left in this city, of course."

How does Josh feel about news media? I showed him my map of about thirty news media outlets and magazines on my iPad and asked for his immediate thoughts. He responded with extended silence and then politely explained why he felt negatively about most of them.

> KRAUSHAAR: My opinion about the media, and I read a lot, is not very positive because I don't feel that it is honest. I remember how *The New York Times* spurred up the energy for war in Iraq, amplifying war fever. That's true of most of these liberal papers, except for maybe *The Guardian*.
>
> AUTHOR: Which ones are you reading daily?
>
> KRAUSHAAR: A lot of the news I get is from local papers, and I read *The New York Times* for broader strokes, but I don't have a coherent media diet. A lot of what I know, and I think this is a problem, I do get off of Twitter or from being active in local politics. I have strong connections to what's going on. I get a lot from other activists from Slack and Signal, which is a like a group [encrypted] text app. I get a lot from these online spaces and hearsay. It's not the best way to get the news, but I don't think there's any comprehensive alternative to that.
>
> I don't think that the American Left would exist in the same way without Twitter because it became a place for sharing coherent leftist narratives that weren't covered in the media. And it's a reason for the generational division.... I think that the leftist millennial Twitter sphere has more literacy and can more clearly see why institutions are failing in capitalism. The boomers basically don't want to think that the things they have been told all their life was a lie.
>
> AUTHOR: How about other social media platforms?
>
> KRAUSHAAR: I avoid Facebook at all costs. I sometimes watch food videos on TikTok. I don't use Snapchat. LinkedIn is gotta be the most psychotic place ever to get your news.[25]

The Visual Director

The electoral campaigns have a visual design director who is responsible for the overall visual identity of the campaign. The director creates posters, handouts, T-shirts, stickers, and social media explainers. They also create social media templates for the campaign's communications director. In the Alexis campaign, the communications director was busy producing updates and testimonials by campaigners. With more resources, the team could have developed more strategic efforts to activate publics across the fragmented and polarized media landscape. The "Public Power" campaign demonstrated the potential of a more developed public relations effort (see chapter 9).

Some of the most prolific designers in NYC-DSA's electoral campaigns are Andrea Guinn and Ronin Wood. They spearheaded DSA's National Design Committee. Ronin explains that the two have collaborated and developed a mutual understanding of style and responsibilities. When Andrea was too busy managing an electoral campaign at one point, Ronin took over some of her design responsibilities.

Ronin was the design director for the Alexis campaign. He studied graphic design and mostly does websites and Facebook ads for companies. He says DSA gives him the opportunity to work on a wider range of genres and communication channels. He loves to create physical posters and see them in his neighborhood, not least with a political agenda that he shares. Having worked for a campaign organized by the Democratic Party, he appreciates how NYC-DSA is using visual design to create political consciousness and stimulate grassroots participation.

Ronin explains that he developed the campaign's visual identity through conversations with Alexis's family and friends. This resulted in a logo with a black fist holding a red rose and a color profile with tan as the main element. Red and black are used as accents. Alexis is Black, and Ronin talks about the revival of red and the red rose as symbols of the democratic socialist revival (see chapter 4). Ronin also uses red to represent the Haitian diaspora in Flatbush. He uses blue as a filler. "Blue is when you've run out of colors and you need a pop"—but not too much because it is associated with conservative patriotism and centrist Democrats. Some of the campaign visuals feature four languages to accommodate the different language groups in the neighborhood.[26]

* * *

In conclusion, the people in the Alexis campaign belonged to a Brooklyn demographic of downwardly mobile young people with culture-driven life choices and flexible family and work arrangements. Their cultural resources help explain NYC-DSA's bold and holistic approach to political organizing, a point that will be detailed in upcoming chapters. The individual-level explorations show that people in NYC-DSA have different levels of experience in the political field, and some are more climate focused than others. Also, there is significant variation in the extent to which participants are motivated by personal grievances. Some have stable and comfortable living conditions and mostly join the organization because of their moral vision for society. There is similarly great variety in the organization's role in the social life of participants. It has a central role for working group organizers, while this is less true for the casual canvassers.

7

Organization and Leadership in NYC-DSA

Leading in social movements requires learning to manage the core tensions at the heart of what theologian Walter Brueggemann calls the "prophetic imagination." . . . A deep desire for change must be coupled with the capacity to make change. Structures must be created that create the space within which growth, creativity, and action can flourish.
—Marshall Ganz (2010)

This chapter develops the analysis of the overall organizational culture of the new NYC-DSA. It offers a general history of the organization's culture and structure, including the leadership approach. This analysis is nuanced through individual-level explorations of perspectives among top-tier leaders. I focus on three leaders who were involved in the development of the Alexis campaign and who canvassed for it. These individual-level explorations conclude the discussion initiated in chapter 6.

A note on terminology: I use the phrase "top-tier leader" as a shorthand for "people with a top-tier leadership role." The leaders continue also to have a rank-and-file role in canvassing and do not sequester themselves on a separate, elevated plane. Leadership status is defined by the role in the collective and must not be exercised in individual, coercive, or hierarchical terms. There is a strong aversion to patronizing leadership.

The primary challenges stem from the reliance on amateurs and free labor. Leaders generally do not have professional leadership experience, and they do not have support from professional administrative staff. The engagement in electoral politics helps overcome some of these challenges, because electoral campaigns have funds to hire staff, and elected officials have access to government-funded staff resources. Also, NYC-DSA has only three salaried staffers. Two of these were added in 2025, and this expansion is crucial for developing a stronger, more professionally resourced organization. The organization is thus based on the free labor of members, who also pay for the operating costs. It is financed by membership fees and donations. The movement leadership approach to some extent contributes to burnout. Organizers are socially rewarded for their commitment, creating a culture of overwork. That is why some top-tier leaders step down after a few years, often shortly after they have begun to master their role.

The top-tier leaders are members or chairs of leadership committees. They are democratically elected as representatives of neighborhood branches, so they are chosen by members of those branches. All members can vote and automatically receive a ballot. The candidate nomination process is open to all members. Candidates must write a statement on their experience, values, and goals.

There are leadership committees at the branch (neighborhood), chapter (city), and national levels. Thematic working groups are organized at the city and national levels. A so-called distributed leadership approach dictates that these various committees are fairly coequal. Individual branches and working groups collaborate with and have representation at the city level, for instance. In New York City, to be more concrete, the citywide committees create rules that the neighborhood branches must follow, but they only create rules if all branches agree to them. The role of the citywide committees, moreover, is not primarily to create or enforce rules for the branches but to create infrastructure for them, do more of the overall strategic work, and serve as an in-house think tank. The citywide committees are formally responsible for the overall political direction and administration of the organization. The national level plays a smaller role than the city level because DSA is so decentralized nationally. The movement party culture in NYC-DSA, for instance, is largely a local phenomenon, although the National Political Committee shares the commitment to combining movement and institutional politics and the strategy of primarying Democratic incumbents in left-leaning districts.[1]

A History of NYC-DSA's Goals and Structure

1982–2016: A Network of Social Critics

DSA was created in New York City in 1982 from the merger of two intellectual political organizations that had emerged in the 1970s as alternatives to the far Left: the Democratic Socialist Organizing Committee and the New American Movement. The new organization was independent of political parties, but it was born with the intention of becoming a faction in the Democratic Party that could move the needle to the left. Michael Harrington, the founder, eschewed sectarianism and argued that the Left could be more constructive by working within the Democratic Party. DSA adopted the so-called realignment approach, in which socialists entered the Democratic Party's organization, such as by participating in state and national conventions and campaign committees, with the goal of transforming the party in the long term. The phrase "left of the possible" came to represent this pragmatic and cooperative approach. This perspective gained traction in the wake of Ronald Reagan's election, when the Left experienced a major crisis in the national political environment. The situation after Trump's election was somewhat similar, except with the crucial difference that the Democratic Party has moved so far right that democratic socialists no longer feel that the realignment approach can be meaningful.

The first cochairs of DSA, Michael Harrington and Barbara Ehrenreich, came from each of the two formative organizations. The differences between these two groups surfaced in personal tensions between the two leaders, with Ehrenreich being further to the left than Harrington and pushing a feminist agenda together with Gloria Steinem. Ehrenreich and Steinem were influential and visible public figures, but Harrington remained DSA's chief spokesperson. The women in DSA have been consistently sidelined in the official narrative promoted by the organization's men. Some of my informants knew Ehrenreich's and Steinem's writings well, but they did not know that they had played a key role in DSA. Few knew that Ehrenreich had been cochair. This indicates that the organization has not yet fully reckoned with its patriarchal past. Women have long been repressed in the socialist movement and therefore organized more in the feminist movement. This pattern dates back to the suffragette movement and International Women's Day in 1908, when fifteen thousand socialist women marched in New York City. The pattern evolved in the second wave of feminism, which took the form of a socialist feminist public sphere with its own journals, bookstores, publishing companies, conferences, academic programs, and festivals.[2]

"The old DSA" (1982–circa 2013) was essentially a political club. Membership was generally around five thousand, and about one hundred to two hundred members were actively engaged. The focus was on national politics and shaping public opinion through writing. The organization did not organize collective action or electoral campaigns. Thus, the organizational design could be simple: each city had one branch and one leadership body, the Steering Committee.³

Neal Meyer joined NYC-DSA in 2012, and his memory of the National Political Committee meetings is that they resembled the amateur political clubs of the 1950s. Members congregated primarily to confirm their political beliefs and share perspectives on national politics:

> I started going to the quarterly meetings of DSA's National Political Committee [NPC] in our cramped little office a couple blocks from Wall Street. In retrospect, these were pretty funny events. DSA's "old guard" dominated the body, and the organization was very weak at the time. So the meetings brought together a lot of comrades who joined in the 1970s and '80s. Small talk focused on complaining about the quality of the bagels bought for breakfast, talking about the public access TV station a member ran in their college town, and describing in excruciating detail how a chapter sold wine at a booth at the annual Labor Day picnic to keep themselves solvent (I have no idea what expenses that chapter had that might have posed such a problem).
>
> At one meeting we learned that a comrade had printed out the sign-up sheets provided by national DSA to table at a nearby campus, but they had crossed out the words "Democratic Socialists of America" and replaced them with the title "Progressives of California" because they thought the "s-word" was too controversial. . . .
>
> The younger members in the room would plead with the older members to acknowledge that DSA's Geocities-aesthetic website at the time was an embarrassment, and that few people would join the org if we didn't redo it. It took a lot of work, but they were finally convinced. . . .
>
> Mostly I thought the meetings were pretty boring, though. What really ground my gears was the regular "lay of the land" political discussion that the NPC would invest a lot of energy into. They'd sit and discuss the political situation in the US for hours: the strategies of various figures in the labor movement, the Obama administration's latest maneuvers, GOP plans for the midterms, what slogans and demands were most popular among the (mostly student-based) really existing left.⁴

"The old DSA" was defined by a high concentration of intellectual workers. Some of the leaders were professors at the Ivy League schools from which many in the new NYC-DSA would emerge. However, they had more privileged jobs and were older when they developed the organization. In 1982, Harrington was fifty-four, Ehrenreich forty-one. Other people in leadership included Irving Howe and Meyer Shapiro, who were sixty-two and seventy-eight, respectively. A few were around thirty, including Cornel West and Adolph Reed Jr., and they, too, had Ivy League careers. Several leaders of the old DSA had been involved in the Left since the 1950s and experienced one cycle of boom and bust of the Left. They had created or were deeply involved with magazines, such as *Dissent* and *The Nation*. They were an intellectual-activist political class.

The new DSA, by contrast, was a place of downwardly mobile graduates in their twenties, alienated by neoliberalism and struggling to find employment and pay the rent. The people in the new NYC-DSA felt that the methods of the old DSA were insufficient. It is illustrative that *Jacobin* helped start local reading groups that became DSA chapters but that these chapters are not led by *Jacobin* editors. The people in DSA and NYC-DSA leadership are generally not editors, professors, or public intellectuals. There are few academics in leadership, and they are not full professors at Ivy League universities. DSA leaders do not belong to the top tier of powerful institutions and are critical of them.

Describing the old DSA simply as a privileged East Village intellectual club, however, would be incorrect and obscure important continuities with the new DSA. Ehrenreich and Howe had working-class backgrounds. Some were committed activists and had organized at some point. Reed organized with poor Blacks and antiwar soldiers in the late 1960s, for instance. Many volunteered for Jesse Jackson's 1984 presidential campaign. Additionally, the relationship between writing and activism is more complex than one might think. The stereotype of the leftist philosopher does not fit the old DSA. Harrington and Ehrenreich wrote some of the most influential working-class reporting of their respective lifetimes. Harrington's *The Other America* of 1962 is an original analysis of how mass poverty was made invisible to the US public during this era of growing but unevenly distributed prosperity.

Ehrenreich's *Nickel and Dimed* of 2001 was based on fieldwork as a minimum-wage service worker. She used many book talks to advocate for raising wages.[5]

2016–Present: A Movement Party

The first Sanders presidential primary approximated the realignment approach before ultimately illustrating its inherent risk of marginalization. It was an organized faction that formulated an alternative to the Democratic Party's political program. This resonated with DSA, which had abandoned the original realignment approach by the early 2010s. The party had changed and now had a reputation as "the graveyard of social movements." Creating a new party was not feasible, so local DSA chapters took inspiration from the populist insurgent approach demonstrated by Sanders and developed it into local independent organizing, using only the party's ballot.[6]

The major national events in 2016 moved millennials into DSA, creating a new age and class demographic in the organization. New leftist media motivated interest in democratic socialism and DSA but did not guide political organizing. The focus on local collective action in the broader anti-Trump movement explains why local variations became more pronounced in DSA. The new NYC-DSA emerged as a local movement formation. It was no longer simply a public. Whereas the old NYC-DSA had an institutional base in national leftist media, the institutional base of the new NYC-DSA was in campaigns on the ground. Its leadership was focused on institutionalizing the democratic socialist movement and influencing the state government. Thus, NYC-DSA became framed by its new relationship to the state government. In short, the transformation was ontological, spatial, and institutional.

The new NYC-DSA is defined by the focus on collective action in electoral and issue campaigns from 2017 onward. The basic campaign model was in place in 2018, and the same movement leadership approach was used in both types of campaigns. Meanwhile, organizational structures evolved. A key point in what follows is that the new organizational structures have both short- and long-term implications. The electoral work has had wide-ranging implications for the organization's culture and has led to ongoing party building. It has brought the organization into the institutional framework of parties and government. The organization is in the process of clarifying its position in this field, and it is still evolving as a movement party. The electoral work creates expectations for a party identity and shapes the organization's horizon and identity among New York's many political organizations. It further has an organizational and economic function: Campaigns create employment for organizers and serve as a mobilizing structure for the organization.

Significant organizational developments in NYC-DSA since 2016 include the following:

1. The creation of a branch in Brooklyn in early 2016, when the organization became populated by millennials living in Brooklyn between Williamsburg and Bushwick. The expansion in Brooklyn further led to the partition in June 2017 into a North Brooklyn branch (Williamsburg and Bushwick) and a Central Brooklyn branch (Crown Heights, Clinton Hill, Fort Greene). That same year, branches were created in Upper Manhattan, the Bronx, and Queens. The branches are the platforms for local networks around the city and the electoral campaigns that are structured by electoral districts.
2. The creation of the Brooklyn Electoral Working Group in December 2016 with the aim of organizing electoral campaigns. The proposal for the group stated that Sanders had demonstrated popular interest in a new type of politics—democratic socialism—and that NYC-DSA should now build "the electoral wing of a grassroots democratic socialist movement."[7] This formulation signals synthesis of movement and party elements. It imagines government as an institutional base of the movement. The movement is the overarching framework. This chapter details how organizational structures have been developed to bridge movement and institutional contexts. Also, synergies are developed between electoral and issue campaigns, with electoral campaigns evolving from issue campaigns and "electoralizing" movement issues. Issue campaigns have helped build pressure on Democratic leaders from the outside during the annual budget negotiations in state government.[8]
3. The creation of the Citywide Leadership Committee in 2017 to expand democratic grassroots leadership. The Steering Committee

remains and has a conventional outlook of a political organization, with two cochairs and one representative from each branch. The Steering Committee is responsible for administration and developing political strategy proposals. The Leadership Committee is a broader forum of deliberation, with up to thirty members. It can endorse and challenge proposals from the Steering Committee.
4. The creation of thematic working groups in 2017 that soon began organizing single-issue campaigns. The most influential are the Housing, Health Care, and Ecosocialist Working Groups. The thematic working groups are citywide. The labor group (named the Labor Branch) remains peripheral to the electoral work and focuses on union organizing as a form of base building. The Ecosocialist Working Group is central to this book and is discussed in detail in chapter 9.[9]
5. The creation of new caucuses that define political tendencies. The caucuses organize political factions. They are spaces for deliberation and develop proposals for the organization. They send delegates to the annual national meetings. But they are not leadership bodies. The two largest caucuses are Bread & Roses and Socialist Majority. The former was created in the early 2010s and has a Marxist focus on class struggle, particularly the mobilization of the "not yet politically active working-class majority." The latter was created in early 2019 and represents the new practical focus on electoral work and a more reformist position closer to Sanders's. Also represented on the National Political Committee is the Red Star Caucus, which promotes revolutionary socialism. There are extreme-Left caucuses, but they are small, have little influence, and are generally residues of sectarian organizations that collapsed when DSA became the new platform for the urban Left.[10]
6. The creation of the political action committee DSA for the Many in 2020. This multicandidate committee made it possible for NYC-DSA to collect and distribute funds to electoral campaigns, even though it is not a political party.

Distributed Community Campaign Organizing: The Ganz Version

It is hard to fully grasp the culture of grassroots democratic organizations without firsthand experience. Textbooks on management and leadership offer little help. They tend to have a chapter on volunteer management, which is not quite the same. A 2010 volume motivated by the need for "more and better leadership" in world societies included one chapter on movement leadership. The editors nonetheless presented the chapter as an auxiliary and buried it late in the book.[11] Without firsthand experience, our understanding of grassroots movements relies on media sources and occasional encounters. As a child, for instance, I visited hippie communes a few times with my parents, but such experiences are obviously not enough to understand this phenomenon. There are plenty of romanticized images of collective action protest, nurtured by utopian desires and a fascination with heroic and charismatic leaders. They tend to reproduce the David versus Goliath story from the Bible, in which a young shepherd kills the evil giant with a sling and eventually becomes king. Think of Hollywood movies such as *Spartacus*, *Malcolm X*, *Gandhi*, and *Selma*. There are also negative images portraying movements as dysfunctional and sectarian. This includes fascination with far-Left radicalism in *The Weather Underground* and the lifestyles of movements, as in the hippie movies with a tragic ending, such as *A Walk on the Moon*. The production and interpretation of popular culture depictions of movements are shaped by political interests. Leftists have been repressed in film and television through anticommunist blacklisting since the late 1940s. Movements contest dominant norms in society, and this affects thinking about their organizational culture.[12]

Many social scientists were initially dismissive of the new participatory democratic organizations in the 1960s, portraying them as a kind of romantic primitivism.[13] But these movements demonstrated potentials of democratic decision-making and provided new normative frameworks for deliberation.

The situation for grassroots movement organizations changed in the late twentieth century, when the broader field of civil society organizations became professionalized and instrumentalized. Organizations in this field began implementing styles of corporate management under the influence of neoliberalism, bureaucratic distribution of welfare services, and the mediatization of social life (see chapter 6). The corporatization led to the formation of the so-called NGO-industrial complex, defined by a hierarchical division of labor between technocratic management and volunteers. Theda Skocpol describes a shift from membership to management in civil society, as a result of a transition away from nationally federated and locally grounded civic groups toward centralized national organizations.[14] Organizations began relying on professionally outsourced canvassers and instrumentalized collective action to the point where members are confined to generic labor and armchair activism, such as signing digital petitions, sending letters to elected officials, and liking social media posts. This led to a gap between these centralized national organizations and the everyday local experience of ordinary citizens. In some movement organizations, there was a sense that the organization's founding values and goals were compromised by these organizational changes, with leaders who were not baked into the movement. Dana Fisher showed in her book *Activism, Inc.* of 2006 that this transformation in civil society organizations extended into the Democratic Party. The party was increasingly relying on professional consultants and paid canvassers, with no active involvement of members. Institutional theorists might explain this situation in terms of an adaption to broader developments in modern organizational life, with civil society organizations and the Democratic Party recruiting from the same elite institutions and receiving funds from the same donors.

One presidential campaign in 2008 diverted from the general trend and inspired a revival of interest in grassroots *campaign* organizing (not the same as long-term *movement* organizing). The Obama campaign attracted media attention to its organizing approach because the movement aspect was strong and helped lead to victory. Obama talked publicly about his experience as the director of a church-based community organization on Chicago's South Side in the 1980s. But the campaign approach was based in oral tradition, and there was no textbook; so researchers began codifying the approach. Marshall Ganz, a key adviser to the Obama campaign, began synthesizing elements of movement leadership into a framework after Obama's victory.

Ganz brought experience and knowledge of organizing to the Obama campaign and codified it afterward. He does not come from classic social movement studies or use this literature as his primary framework. His thinking develops from his experience as a practitioner helping large organizations create engagement and become more efficient. His main contribution lies in the development of a clear and systematic approach to leadership in nationwide campaigns. Ganz has an administrative mind more than a political one. He does not address how citizens can be involved in the political process. Ganz worked for sixteen years as a staffer for United Farm Workers. In his youth, he had worked in the civil rights movement.

Ganz completed a doctorate in sociology in 2000 on the unionization of farmworkers in California and began teaching leadership at the Kennedy School of Government at Harvard. This helps explain his attention to social relations in organizing. His framework in the 2010 chapter "Leading Change: Leadership, Organization, and Social Movements" also draws on positive psychology to explain the power of hope and how it can be leveraged. The chapter focuses on leadership, motivational and persuasive storytelling, and the efficacy of movements to train leadership and be strategic. Further, he explores how movement organizations can catalyze action through hard asks and formulating specific outcomes with deadlines. He focuses on the campaign as a motivational tool and a strategic effort with specific objectives. The framework works from philosophical aspects of intentional relations to the more practical aspects of strategy and planning. The framework draws from research literatures, but it is formulated as a kind of advanced recipe, a prescriptive conceptual outline, based on experience by an expert scholar-practitioner. It is not primarily an analysis of arguments in the literature but rather uses arguments to support and refine his practical knowledge. It can thus be understood as a form of design methodology.

Ganz's framework was popularized in 2014 with the publication of a free digital booklet titled *Organizing: People, Power, Change*. This booklet focuses more narrowly on campaign organizing. It is less conceptual and historical, but there was not much history in the framework in the first place. Ganz's chapter in the 2010 handbook does not address the concepts of democracy and citizenship or refer to scholars of grassroots democracy such as Alinsky and Stout. Ganz published a monograph on the topic in 2024.

The 2014 booklet is written by two young community organizers, Shea Sinnott and Peter Gibbs. The authors wrote from the perspective of managing and facilitating volunteers, not political leadership. The operational-technocratic interest is evident in the formulation that the booklet is intended as a "guide" for "effective community organizing." Movement ideas are extracted from their context of origin into a context-free model. The booklet circulates as a free PDF file and thus brings the ideas—in a refracted form—into new contexts. The model can serve as an instrument for campaigns within corporate structures, but it can also be read as a particular summary of ideas that guided the organizers of the Obama campaign, including Tascha Van Auken, who brought this approach into NYC-DSA a decade later.[15]

The booklet adapts Ganz's framework into a series of instructions, with brief explanations of the underlying principles. The potential of the framework is to recruit volunteers and turn them into committed campaigners. This primarily happens when organizers develop intentional relationships with volunteers around shared goals. Organizers help make volunteers see and feel how their own interests align with the campaign. Organizers further work with the expansive principle of developing volunteers into organizers. A team of five volunteers can in theory become five teams. This is important because of the generally high turnover of volunteers. The principle of equal and intentional relationships extends to the more general level of the campaign organization as a whole: Leadership is shared among teams that are coequal but serve different functions. This is why the model is called the "snowflake" or "distributed leadership" organizing model. A central leadership team develops strategy and long-term goals for the organization as a whole, while local teams are testing the strategy and providing feedback. There is not a unidirectional hierarchy. Having collective instead of individual leadership also makes the organization less vulnerable to volunteer turnover.

A big part of the Ganz approach is the operational dimension. There are numerous concrete recommendations for how to accomplish the goals, including the following: (1) The individual organizer should build a team by having one-on-one meetings to understand the person's situation and a shared understanding of the goal. A full-time organizer should aim for maintaining relationships with no more than ten volunteers. (2) The individual organizer should create clearly defined roles and responsibilities for volunteers and help them develop an understanding thereof. The organizer creates collective identity and purpose by telling a story of who they are and why they are organizing, why the community is coming together around this challenge, and what they are going to do about it. Volunteers too are encouraged to tell their story and why they are getting involved. No one is assigned a task without dialogue about the purpose and possible ways of contributing. (3) The campaign should have concrete and realistic goals that can be understood as incremental steps toward a larger goal and vision. The overall goal of a movement cannot be achieved in one campaign. The core principle of creating intentional relationships is thus operationalized in a series of concrete actions in an overall progression from building to wielding power in a community. In this process, a community is transforming into a constituency committed to action, and its power is wielded to create change.

The embodied dimension of this organizing approach has importance in a media-intense political culture. I develop this perspective in the following with inspiration from media and communication theory, emphasizing the implications for political agency. The tradition of community organizing from Alinsky to Ganz places central importance on in-person meetings. The key interaction happens through *media of the first degree*: human bodies. Participants are sharing place and time and communicating directly with the body as whole. When people are physically distant and communicate electronically, media can reach across physical distances but not fully overcome the bodily separation. The general perspective here is that embodied communication and experience are essential to building close relationships. The pandemic showed us that electronic communication with family and colleagues lacks affordances for experiencing vital aspects of human connection. What is more, electronic media objectify communication into symbols in visual or auditory material form and distribute them across time and place. The body, for instance, becomes an image that can be watched anywhere on a phone screen. This image can also be manipulated and thus undermine the communicative authority of the individual. The small scale of embodied communication further helps explain why individuals can interact more continuously in the flow as equal agents. It is more difficult to focus a conversation when it involves many participants and electronic distribution. This is especially true of social media platforms. Media of first degree also shape the style of communication and thinking. This includes oral culture elements such as improvised interactions and adaption to a specific local setting. These are important qualities for building trust, engagement, and managing complexity. Tascha Van Auken found that it was harder to get new volunteers to take on big projects and grow as organizers during the pandemic: "It's a lot harder to do that [during the pandemic]. Just developing a relationship with people to the point where they want to give that much, that is a whole piece that's a lot harder when you're not in the same space as people. And we've done it, and we have a really amazing team, some of [whom] I've barely met in person. But it took a lot longer."[16]

The leadership approach codified by Ganz is primarily developed for the organizational microlevel of the individual organizer's in-person interactions with volunteers. The approach is less developed at the organizational macrolevel. It is essentially the art of organizing volunteers into a campaign. NYC-DSA has to some extent built its organizational macrolevel from this microlevel approach. Additionally, the Ganz framework does not emphasize the political agency of participants, which is so important in a movement party context. In the version popularized by Sinnott and Gibbs, the campaign has a predefined goal. The volunteers are not defining that goal as participants with political agency, and they have no influence on the outlook of the overall organization. The approach does not organize people into the development of a political vision or a political organization. It does not include processes of deliberation and decision-making in political organizations with the aim of developing solutions to complex problems with a more long-term perspective than the individual campaign.

The remainder of this chapter focuses on the organizational macrolevel in NYC-DSA and attends to the ways that it is shaped by the tradition of movement leadership codified by Ganz.

The Role of National DSA Leadership

DSA is decentralized, but the national leadership creates some infrastructural elements for local chapters. It defines principles for electoral and pressure campaigns and channels resources into campaigns that it grants the status of national priority campaigns. The idea that leadership committees have a coordinating function is emphasized at both the local and national levels, much in line with the distributed leadership principle.

The National Political Committee (NPC) assumes a public voice for the organization by issuing statements on political positions. The national level of the organization, however, is to some extent an administrative support unit for the local chapters, adopting a coordinating role and creating infrastructure for them. This includes the role of the national director—a role filled by Maria Svart from 2011 to 2024. The national level of the organization has more importance to smaller chapters without the critical mass of NYC-DSA. Overall, however, it is not the central space from which ideas, strategies, and resources are defined and distributed. The centers of action are the local urban worlds of the movement. For leaders in NYC-DSA, serving on national leadership is considered an act of solidarity, albeit with some degree of prestige.

DSA's constitution has somewhat functional and abstract language. It lists information. By contrast, individual campaign proposals, such as the 2022 Tax the Rich 2.0 proposal, adopt a more situated perspective and are more argumentative and passionate. This reflects that the national leadership is a somewhat synthetic body and that movement energy is leveraged locally.

The relative absence of models of organizational macrolevel leadership creates challenges for a multitendency organization such as DSA and its development as a movement party (see chapter 8). Internal tensions and fragmentation evolved when DSA expanded in 2016. DSA became a sort of umbrella organization for a wider range of factions on the Left. The low barriers to entry meant that far-Left networks could join the organization without actually sharing its values. This is called "entryism." Also, people with little experience in political administration and leadership could rise to national leadership quickly. These are some of the reasons the NPC has been internally divided, while the campaign-based organizing in NYC-DSA has helped make this chapter more unified on the ground.

The NPC has sometimes struggled to respond to crisis situations, resulting in skepticism about the committee's legitimacy. This skepticism has evolved into toxic conflicts in DSA's national "discussion board," which was paradoxically launched by national DSA in 2018 to strengthen internal communication at the national level. Among my informants, even movement pundits had left the forum for this reason. People who only follow the organization via media communications are easily misled to think that the toxic atmosphere in the forum is representative of the organization. Social media platforms allow members to communicate directly and spontaneously, and this presents new challenges to grassroots organizations.[17]

The desire for a clearer national party identity can also be registered among the organization's graphic designers. Andrea Guinn and Ronin Wood spearheaded the National Design Committee. They believe that "a more unified visual identity for NYC-DSA" and for DSA nationally would be helpful. The organization's visual identity is inconsistent in large part due to the turnover of designers working separately on individual campaigns. The organization does not have staff designers, as the Democratic Party organization does. Nonetheless, Guinn and Wood have argued against having the same brand template for all visuals, an idea that was part of a 2022 proposal to develop the party dimension (see chapter 8). They instead recommend that core elements be repeated.

A practical challenge to leadership at both the local and national levels is that serving on a leadership committee is a demanding "second job." Until 2023, NPC members were unpaid, but the $2,000 monthly stipend is so low that they still need a day job to get by, especially if they live in a big city. Their leadership job is less precarious now but is still precarious. Some organizers and top-tier leaders burn out after a few years. At a meeting about challenges to DSA's electoral work in May 2022, Assembly Member Zohran Mamdani expressed concern about this problem: "How sustainable is it that our organization is based on volunteer work on this scale?" The most common model of a sustainable work life for top-tier leaders based in New York is to have a paid job in a DSA-backed electoral campaign or to be a staffer for a DSA-backed elected official.[18]

The party dimension of NYC-DSA transcends the realm of a strict grassroots movement project and requires expertise. The people with top-tier leadership responsibilities have educational experience and intellectual resources to understand formal institutions such as government and to communicate ideas. Although they do not have broad foundations in political administration, they collaborate with a range of elected officials on climate legislation (see chapter 9) and are not locked into an affective state of acting from indignation. NYC-DSA represents a new development of leftist grassroots movement toward institutional politics.

Leadership in NYC-DSA

The campaign community organizing approach has thoroughly shaped the organizational culture of the new NYC-DSA. The campaigns are the central structuring activity and have helped implement the principles of intentional relationships and clear and concrete goals throughout the organization. The organization has also adopted the principle of managing teams, in this case committees, around tasks rather than hierarchy. Tascha Van Auken had internalized these principles from working on the Obama campaign.

> I'm interested in organizing the grassroots in a way that typically is done more professionally on campaigns. The Obama campaign was a perfect example. It was so well-run, and set a really high organizing bar for me. But all of that power went away and was killed by the Democratic Party. So, without really knowing it, I was always interested in how to create that in a grassroots setting that gave power to people doing the work rather than a couple of people running a campaign or running an organization.
>
> I think that there are a lot of preconceived notions that people have about elections and how you're supposed to work on an election. A lot of those have to do with ideas around power and who's more senior and all this crap that I personally don't like and is a big reason why a lot of people don't like to be around political campaigns.
>
> I don't think we reinvented the wheel or anything. I think other organizations have done very similar things. But, for ourselves, it let us reimagine what working on a campaign is like and create a very noncompetitive and friendly culture where we're training each other to learn how to do these things better. If you are a volunteer, one day you're going to be asked to lead a canvas. And then once you lead a canvas, you're

going to be asked to identify other people to lead a canvas and to train them. And your success is viewed as, how many people can you train and elevate around you, rather than how important you, individually, are.[19]

In an attempt to understand this microlevel organizing better, I asked Van Auken how dysfunctional tendencies can be handled. The absence of formal hierarchy seemed like a potential weakness to me. At the same time, I experienced the robustness of the Alexis campaign. She replied that disciplining happens but that constructive organizing minimizes the need for it:

> It can just take one person to really suck the life out of an organizing space. The important thing to do to prevent people like that from taking over is to have very specific goals, structures, processes, so that, like, if in a meeting somebody is trying to take over, you can just be like, "You are breaking this contract we have with each other. You're clearly just trying to disrupt it." The other thing is just people always having goals and roles and the project being about "We're working together to do this thing, and here are the things we're doing." When everybody's really focused on work, it becomes really clear when you have somebody under that space who's not. If you have a dysfunctional place, where it's not clear what the project is, and there's, like, weird power dynamics, and there's not a clear structure, then it's very easy for people to come in and take it over. My central principle is it should be very easy to volunteer, and it should be very clear what to do.[20]

In my field experience, the NYC-DSA members who expressed hopelessness and negativity about the organization and political life more broadly were the members who did not canvass or organize. They were known as "paper members."

The Selection of Leaders

Organizational socialization is a key concept in organizational theory. It assumes that recruitment and onboarding processes are important to reproducing the organization's values. In NYC-DSA, leaders are recruited among internal organizers who have been socialized into the organization, usually for more than one year. People build and demonstrate their leadership potential through organizing in a working group, committee, or campaign. Aina Lakha, one of the most experienced citywide leaders, explains,

> Everything hinges on like, you know, we run these big field operations, Tax the Rich and the electoral campaigns. We're trying to find people who want to take responsibility for something and who can support and take responsibility in executing something. We want people to step up, get involved, carry out a project, learn how to do that, learn how to be a responsible leader who moves other people. We build our leadership pool out of those people, people who were involved in campaigns. Now, some of those people come with ideological training and backgrounds of their own; some of them don't. Along the way, they might pick it up. Usually, people who become leaders are people who develop some intellectual interest, either like a book interest or they're just interested in how politics works. And what's happening is you talk to people, and they learn a lot through that. It's a very political environment.[21]

Here, Lakha expands on the task-oriented approach and adds political intentionality to the picture. Although leaders do not have to be politically educated when they enter the organization, they generally have some level of political knowledge to be elected to leadership, and they can rise faster if they have political expertise. Leaders are expected to have skills to strategize and organize deliberative and decision-making processes around the organization's political goals. Leaders in NYC-DSA are political leaders and not campaign managers, after all.

My informants were committed to the organization's diversification and proud of the leaders and political candidates representing this effort. The outlook of the current top-tier leaders speaks for itself: They are all in their late twenties or thirties and diverse with regard to ethnicity, gender, and race. They are all downwardly middle class, broadly defined. Most have master's degrees, some have PhDs, and most are struggling to maintain a living wage in the cultural, media, or NGO fields; some are electoral campaign managers in DSA or staff members for DSA-backed electeds. With regard to cultural style, they follow the postbohemian style found among the Alexis campaigner (see chapter 6).

There are also certain personal qualities that the organization values. There is an expectation that leaders, candidates, and elected officials are authentic and collectively minded "movement people" who continue to canvass and above all be collaborative, egalitarian, and respectful to stimulate trust and create a positive experience of participation. This is what I experienced in the Alexis campaign. Additionally, a leader should be able to drive campaigns and develop new organizers. The collectivist ideal can further be defined against the figure of the ego-driven leader who seeks personal power and adulation. None of the leaders or elected officials cultivate a celebrity persona. Toxic behaviors are not tolerated. These attributes are important because leaders in NYC-DSA are tasked with conflict mediation among grassroots organizers. The community organizing approach is a big part of the explanation for why the conflicts in NYC-DSA have not been so serious that they have made the organization dysfunctional.

The internal recruitment of leaders helps create leaders who understand the organization's values and earn trust. Leaders are not recruited based on a generic profile or coming directly from a field of similar organizations. This limits the talent pool and creates a somewhat insular and idiosyncratic organizational culture. NYC-DSA could gain access to wider flows of resources by hiring professional NGO leaders, but it is not quite aligned with the values of that field. It also does not have the finances to attract such talent. NYC-DSA relies on unpaid leaders with a considerable commitment: Organizing committee members are expected to work five hours per week for a few years; but some of them work more, and some citywide leaders reported working twenty hours per week on average. These aspects partly explain why NYC-DSA has had a shortage of leaders and why campaigns are sometimes understaffed and lack resources for strategic development and implementation. The Alexis campaign had few resources beyond keeping the daily campaign operations running, and the "Public Power" campaign was exhausted when BPRA passed in 2023 (see chapter 9).

Organizational-Political Consequentiality

NYC-DSA expands on a democratic socialist tradition of explicitly combining ideology and organization, seeing democracy as both means and end. In this respect, NYC-DSA's approach contrasts with popular neoliberal approaches to political organizing that elide ideology.

NYC-DSA leaders see grassroots democracy and ownership as necessary means. The goal of a campaign is not just to win constituents over but to democratize the political process. The leaders argue that organizational independence from the Democratic Party is necessary for advancing this goal and enabling a political project that is critical of elites. The party will not support this project because it is funded by elites. NYC-DSA's approach to political organizing thus incorporates a political economic perspective with historical roots in democratic socialism.

Whereas the Ganz framework focuses on progression in the mobilization strategy of a campaign, NYC-DSA focuses more on political strategy and the political project of the organization and movement as whole. NYC-DSA leaders apply principles popularized by Ganz to political leadership, such as operational rigor and purposeful action. This shows in how they integrate ideology and organization in their understanding of strategy and practice. I discovered an emphasis on organizational-political consequentiality when organizers and leaders described their experience of entering NYC-DSA. They would typically describe this aspect of the organizational culture without using a specific term for it. They experienced a culture of constructive conversations about substantial political goals and rigorous operationalization. The leaders contrasted this culture with past experiences of dysfunctionality in far-Left organizations or in the NGO-industrial complex.

Tascha Van Auken and Jack Gross both used the word "serious" to describe the organizational culture, though Jack struggled with the word because he thinks that it could misleadingly convey arrogance. This touches on a deeper ambiguity in the organization between the ideal of popular democracy and the urban movement culture.

> VAN AUKEN: The thing that I really have loved about DSA from the moment I joined was just like, "Oh, this is a serious space. This is serious." Not trying to dilute ourselves or take on things that are fanciful, more performative, but trying to really figure out power and how to build power.[22]

GROSS: When I joined DSA and learned how everyone worked even under the challenging conditions of the pandemic, I realized that out of all of the projects I'm involved in, the DSA project people are extremely serious about building something that can move the political terrain and lasts for a long time and isn't just about reacting to the moment.

AUTHOR: What do you mean by "serious"?

GROSS: I'm wary of "seriousness." It could sound snooty, or we should be professional or something, which isn't what I mean. But it's sort of looking at the task that you've set out for yourself and taking stock of what it takes to achieve the task you've set out, and then doing all of the steps in between. It's astonishing how few left-wing organizations do this. There's a whole cottage industry of trainings for progressive, left-wing, radical, nonprofit employees and activists, especially in New York City, where you get a chart of strategy, tactics, and media stuff for political pressure. Everyone knows how to do these things, but so few people actually carry them out and build the necessary structures.

Tascha was instrumental in building this organizational culture, along with other people who built the Electoral Working Group in Brooklyn, such as Sam Lewis, Michael Kinnucan, and Cea Weaver. What they did was look at the task: "We want to change the political terrain in New York, and we want to do that on the level of the state legislature, which controls the levers of so many people's lives. Basic material conditions are truly controlled by this group of people in Albany." And then they're like, "How do you win elections?" They find people who know how to do that, like Tascha. Learn from all of the failures of the Democratic machine way of operating in New York and the successes and failures of the "new wave of distributed leadership organizing," which was developed in the Obama campaign. "Are these sets of tactics we can use to build a socialist electoral project here? Yeah, some of them are." Then they built that. And that's the seriousness, being sober enough to look at a situation and identify a possible intervention and then smart and thorough enough to figure out the tactics appropriate to that intervention and committed enough to follow through and build the thing.

> In general, people in DSA are serious about their participation in a socialist organization being something that will have real impacts. It's not primarily a social club. It's not primarily a place where you refine a political sensibility or an oppositional attitude to the rest of the world, which is, like, my experience of most left-wing groups. It's primarily a place where people can learn to think of themselves as bearers of possible actual change, like cogs in an organizational machine and leaders in shaping the direction of that machine.[23]

The word "serious" has a couple of meanings here. Van Auken explained in further detail her goal of *taking grassroots seriously* by creating forms of participation defined by authenticity and integrity. This crucially involves consequentiality between grassroots values, political goals, and organizational execution. Taking grassroots seriously, in other words, means integrity all the way, not just in messaging or end goals. Therefore, taking grassroots seriously also means *taking the organization's political goals seriously* and not digressing into mere experiences of identity. Doing so requires persistence and some degree of efficiency. Seriousness also means *being smart and ambitious* by thinking analytically and strategically. This involves rigor, persistence, and consequentiality in the implementation of organizational and political values.

Managing the Relationship Between the Movement and the Electeds

A distinct task for NYC-DSA leadership is managing the relationship between the movement base and its elected officials in state government. After the grassroots have done all the work of recruiting and campaigning for a candidate, the candidate then relocates institutionally when they are elected. The elected official's primary base of operation is no longer DSA in New York City but the government in Albany. The "movement candidate" becomes a movement politician who works within government structures while collaborating with the movement organization. Getting elected involves organizational decoupling and transformation from movement to government, New York City to Albany, city to state. Elections have helped expand NYC-DSA's horizon beyond the urban movement, and this has proven important to its climate efforts. However, the organizational decoupling also creates challenges. The grassroots participate in the electoral process but not the political process in government. In political parties, politicians are recruited and accountable to the same organization, but this does not mean that parties do not have the challenge of connecting electeds and constituents. DSA electeds have a closer relationship to their constituents than electeds recruited by the Democrat Party, which no longer have a grassroots base (see chapter 5). The nine DSA electeds (out of 219) in state government work together as a team and regularly participate in public events organized by DSA or in other public events alongside the NYC-DSA grassroots. They support new movement candidates and participate in canvasses and demonstrations. The organization helps them sustain the relationship with their constituents and canvass for their reelection. The electeds also collaborate closely with NYC-DSA to develop legislation and leverage the potential of movement campaigns promoting this legislation (see chapter 9).

The first NYC-DSA elected, Julia Salazar, describes her entry into government as a transformation:

> When I assembled a team of staff for the first time, there were positions on my team for organizers, like community organizers—people whose job was not only to provide constituent services but also to provide opportunities for people to lead and be empowered. But when I submitted those titles to the senate, they said, "This isn't a title." We had to change the titles of people on staff. That was one example that really stuck out for me. Sometimes it's more subtle than that. How we use movement language with each other. It isn't only that the legislature in Albany is not a grassroots movement environment. A lot of people in the legislature are not the same generation, and many of them had a more traditional trajectory to becoming politicians. They were a staff for a legislator and then ran for office themselves. They don't speak that language either. So we're just having to sort of switch to be able to communicate our message effectively.[24]

NYC-DSA leadership created a committee in 2020 to strengthen collaboration with electeds. The Socialists in Office Committee (SOC) is a forum of NYC-DSA leaders and electeds. It meets weekly to strengthen alignment and collaborate in mutual interest: The movement can learn from experience inside government, and the electeds can sustain their grassroots base and use NYC-DSA leadership as an advisory board. SOC is composed of Steering Committee members and working group organizers in NYC-DSA and the organization's state senators and assembly members. One of the NYC-DSA leaders, Gustavo Gordillo, speaks to the role and relevance of the SOC: "It was through the Socialist in Office Committee that I realized how much power you have over a politician if you can just provide consistent good advice strategically on policy. The Republican Party has cut public budgets so much that all the government is underfunded. Politicians have no capacity to do research or develop strategy on their own. They totally rely on either lobbyists or, in our case, an organization like DSA."[25]

Clearly, the electeds are not out of the movement orbit, and that is also not a major concern in the movement. The main concern in the movement is that the number of electeds is too small to have significant influence and that the electeds are marginalized by powerful Democrats. Salazar talked about how the negotiations in government are deliberately opaque in favor of the governor's office and the Speaker of the House. Even when the Democrats gained a supermajority in government, the leftists were being marginalized. Even so, Salazar think it is important to have electeds who can communicate what is happening in government to the working class.

> SALAZAR: Everyday people and even people who are involved in electoral campaigns, legislative advocacy, like I have been for years before running for office, we really can't possibly understand the experience of being in the legislature because it is so opaque. It was a steep learning curve. I ran as part of the Democratic Party, even though I'm a democratic socialist, because it's the only way to get elected in New York. The year that I was elected, a slate of progressive grassroots candidates were also elected. They weren't socialist, but they still are and were progressive grassroots candidates also in the party, and at the same time, the Democrats took control of the state senate. That was huge. As a result, everyone around me was also kind of winging it, just learning along with me, even the majority leader, who'd been in Albany for a long time. So we were all learning to-

gether how to operate as a unified Democratic majority. But in terms of some things, you're really not taught anything before experiencing it as a legislator about how bills are actually passed, how committees work. I've studied history. I have studied political science to some degree, although not formally, and volunteered on campaigns and worked as a community organizer advocating for legislation. So I understood the fundamentals of the legislative process, but there are things that are not formalized and that are the reality of how legislation actually gets passed and becomes law: who to talk to, what works and what doesn't, how to persuade people, who are the power players in Albany, not formally but informally.

AUTHOR: Could you give an example?

SALAZAR: There's an expression in Albany when somebody says, "the second floor." This always refers to the governor's office. Maybe we're having a conversation about a certain piece of legislation, say, a climate justice bill, and someone might say, "Well, where's the second floor on it? What is the governor's position? Have you talked to the second floor? What have you heard from the second floor?"—in terms of people leaking information and passing information. Since becoming a legislator, I have really tried to relay information about things like the budget process, which has a profound influence on everybody's lives—and yet it's so opaque. Being able to translate that information to everyday people is really important, especially when we're trying to build a working-class movement.[26]

In conclusion, movement leaders and electeds make NYC-DSA a movement party not only through formal organizational structures but also through collaboration based on shared understandings of the political project and organizational culture as a whole. This understanding among leaders and electeds evolves from interpretations of the national political environment and their experience in individual movement projects. To deepen my account and illustrate the movement party approach in more detail, chapter 9 examines the "Public Power" campaign. This multiyear campaign is one of the major achievements of the new NYC-DSA and its most influential climate effort to date. Next, though, I add nuance to the relationship between ideology and organization by exploring perspectives among three leaders.

Three NYC-DSA Leaders in the Alexis Campaign

Two distinct but overlapping circles of leaders canvassed for the Alexis campaign. The first circle was the leaders from the Central Brooklyn branch and its Electoral Working Group. They had not generally been deeply involved in developing the campaign, but they shared a general commitment to their branch and to electoral work. These branch leaders generally did not canvass every week and acted as a second-tier support network. They included Grace Mausser of the organizing committee and Electoral Working Group members Sam Lewis and Tascha Van Auken.

The other circle of leaders were Citywide Steering and Leadership Committee members. They were more involved and took the lead in a special task force to boost the campaign in its final months. This effort was named the "Door to Ballot" program, and the basic idea was a form of deep and personalized canvassing, with each canvasser focusing on a small number of voters.

Let us consider three of the citywide leaders. They all live in Central Brooklyn north of Prospect Park, are close to thirty years of age, and crossed paths with far-Left organizations before joining DSA. They illustrate and promote the ongoing efforts to diversify the organization.

A snapshot of their backgrounds and trajectories into NYC-DSA is instructive:

- Gustavo Gordillo, Ecosocialist Working Group (ESWG) Steering Committee since 2017, National Political Committee (NPC) from 2021 to 2023, and co-chair of NYC-DSA 2024-2026. Gustavo has a middle-class background and talks about life and politics with ease and confidence. He graduated with a degree in English from an Ivy League school near New York City, where he became sensitized to leftist politics. He joined NYC-DSA in 2017, serving for years as an organizer of ESWG before being elected to top-tier leadership.
- Estafania (Tefa) Galvis, ESWG organizing committee since 2019 and NPC from 2023 to 2025. Tefa has a lower-middle-class background and personal grievances as a woman immigrant of color with precarious labor conditions. Her political perspective is radical. She immigrated to Florida, where she studied theater at the University of South Florida, and in 2018 moved to New York City to be part of NYC-DSA. She organized with ESWG before being elected to the NPC in 2022.

- Aina Lakha, Citywide Leadership Committee (CLC) since 2019 and Citywide Steering Committee (CSC) representative for Central Brooklyn since 2020. Aina has a middle-class background and talks about life and politics with ease and confidence, although they have a complex personality and political identity, with grievances related to gender and race. Aina was involved in the far Left while doing their PhD in literature at Columbia. They joined DSA in 2019, adopted a social democratic perspective, and were elected to top-tier leadership within months. They are one of the longest-sitting leaders in the organization and have managed two electoral campaigns, both of which led to victory, and are now deputy chief of staff for one of the winners.

Gustavo Gordillo is twenty-eight, white Hispanic, and single. He graduated from Yale School of Art and in 2016 moved to New York's Lower East Side, where began working at an art gallery. His political interest was stimulated by his college teachers, but he only became politically active in 2016.

> I had not really been exposed to anything besides capitalism before I took this class called Black Radical Cinemas of Latin America in college. I was in a university that was, like, a finishing school for the ruling class. The films in the classes were mostly by Third World Marxists in the '60s in Cuba, Brazil, Mexico, and Argentina. That was my introduction to socialism. I then worked at art galleries for years, where most workers are downwardly mobile millennials, kind of the same demographic as DSA membership. It wasn't until the 2016 election, and specifically Trump's election, that I thought that radical politics actually could be successful in the US. Some of my friends had been in sectarian leftist organizations in the early 2010s and then joined DSA. Because of them, I joined DSA in 2017.

Gustavo took special interest in climate. He was one of the founders of the ESWG in 2017. He started doing electoral canvassing in spring 2018 for AOC in the Bronx and Queens and for Salazar in Bushwick shortly after. Gustavo moved to Brooklyn, "since most people lived there and that's where the campaigning was happening." The Lower Manhattan branch did not have electoral campaigns.[27]

Gustavo has served on the organizing committee of the ESWG since the beginnings in 2017. He had a central role in the "Public Power" campaign, helped electoralize it in 2022, and managed one of the electoral campaigns that evolved from this, Vanessa Agudelo's campaign for state assembly. Gustavo also helped develop the BPRA bill. Nationally, Gustavo serves on DSA's national campaign commission for a socialist Green New Deal. DSA made this a national priority in 2021.

Gustavo describes his political perspective in analytical terms, thinking as an organizer with an academic background. When the conversation turns to a particular political issue, he explains what solution NYC-DSA can propose and how it can organize to gain influence. He does not dwell on his emotions or personal experiences in politics. Gustavo has the skills to participate in philosophical debates but focuses on developing practical strategy and organizing. His perspective shifted in his transition from academia and the arts to becoming a union electrical worker. He will be a central source on ESWG's history in chapter 9.

Tefa Galvis is thirty years old and grew up in Colombia to "a white farm lord and an Indigenous mother," she says. She is currently the national director of a nonprofit working to eliminate subminimum wages in the service industry.

In 2007, at the age of fifteen, Tefa immigrated to the United States, specifically a low-income Black neighborhood in Tampa, Florida. An early politicizing experience was watching TV coverage of the murder of Trayvon Martin in 2012. Martin had lived nearby in Miami Gardens, a low-income Black neighborhood across the state. "I was shocked and realized that he could have been our neighbor. And this was happening while the KKK was handing out fliers in our neighborhood." This is but one experience that made Tefa more skeptical of mass media: "The news are not regulated enough. I only go to targeted media because capitalism owns the media, and all the media does is make me angry, and I'm already pretty angry from what I'm seeing every day."

Tefa struggled as a low-income immigrant woman to get an education and livable working conditions. She talks about how tuition hikes and student loans exploited her generation of students. She then had a precarious minimum-wage job as a stagehand for six years. Meanwhile,

Tefa did extra work as a political activist and union organizer, which involved additional challenges.

> GALVIS: I was in the Freedom Road Socialist Organization, and we were bringing people in but didn't have a way of building with them. I was like, "This is going nowhere." The people there were unwilling to listen and very dogmatic. In 2015, I was burned out from doing reactive organizing. So I quit.
>
> For the majority of my adult life, I was someone who thrived on crisis. It also has to do with my job and material circumstances. I was a union worker with the International Alliance of Theatrical Stage Employees—crazy hours and never a steady pay for, like, six years. Then I came here [to New York], and I was a waitress and a bartender. I ended up having a major crisis where I had to leave New York and move to Colorado for a while. In Colorado, I helped build a weed farm. During the pandemic, I had a side gig in a music studio. It wasn't really until my material circumstances improved with my current job at One Fair Wage that the situation changed. I am recovering now. I go to therapy now, go for walks, bike, . . . but I'm on the journey of figuring it out.
>
> AUTHOR: How did you get into DSA? What made you stay?
>
> GALVIS: I went to Labor Notes in 2018 [a national conference for union activists] and met people from DSA there. And then when AOC got elected in 2018, I decided that I wanted to continue to do this work, but I needed to do it somewhere where it is going to work. So I moved to New York City in 2019 and became a fully active member of NYC-DSA. I believe in this organization because I believe in organized strategic efforts. As a Marxist, I need a platform to organize people who are disorganized, so that we can actually do something. I believe in civil disobedience protests, but it is important to have a plan—knowing your long-term goals, being strategic about your messaging, knowing what the next step is going to be. What you are gonna get people to do next? Who are gonna be involved? What are the repercussions?

NYC-DSA's approach to organizing helped Tefa forge a more positive path, following more than a decade of hardship. The comradeship and solidarity in DSA helped, too. "DSA is definitely a mutual aid sort of network" that has helped her "have a good life in New York," she says. Tefa reports that "comrades went out of their way" to help when she broke a leg in 2021. She started dating a comrade in 2022.

Tefa gives a concrete sense of how campaigns can take on symbolic significance to the organization's history and the lives of leaders. In the Alexis campaign, Tefa sees her long-term struggles for immigrant rights, racial justice, and climate coming together in one race in a strategic location.

> Why I see it as a high-stakes race, besides Kevin Parker, is that I've experienced firsthand the high level of disrespect that exists within the Democratic Party when I campaigned for Andrew Gillum in Florida. I come here and realize that the Green New Deal is the umbrella issue that we need to fight for, and David understands a Green New Deal. More than that, I'm an immigrant, and immigrants are at the bottom of the workers' scale because a lot of immigrants are undocumented. They're second-class citizens. They barely ever make it out of retention centers. Kids are being starved and abused at the border. If someone like David can win such a hard race in the senate, or even attempt to win, it is already a big drop of hope. We often think that the Republicans are our enemy, but I think our true enemies are the Democrats, and we need to figure out how to get them out of power if we really want to go face to face with the Republicans and the white supremacists. If we are able to do that in Flatbush, which is a working-class immigrant district, we can do a lot of stuff, and the numbers we are having now, those are big things that we're making happen. It's proving the work that we are doing as an organization, strategically connecting our vision to a district—that's what David's campaign is showing. To me, it would be redemption for all of those years of fighting, crying. . . . It's really hard to get up and organize every day, continue this fight, and so winning this campaign with David would be just, "Hell yeah, redemption is all freaking worth it! Let's move on!" It is a very special campaign, a truly working-class organizer campaign.

Here, Tefa is entangled in emotions of hope and the role of electoral campaigns as a mobilizing tool for the movement. Creating the campaign was considered an achievement in itself by some leaders, but this perspective was not without weaknesses that reduced its likelihood of success (see chapter 6). There were leaders who cautioned against being carried away by optimism. This surfaced in a branch meeting about potentially taking on another campaign in spring 2022 on short notice. Andrea Guinn, who was then managing Kristen Gonzalez's successful campaign for state senate, stood up against the room's prevailing consensus. Guinn argued that the organization should not commit to further campaigns that year because it lacked the necessary resources and doing so would put other campaigns at risk.

Politically, Tefa is attached to radical Black socialist thinkers. "Assata Shakur is my ultimate favorite. I read her autobiography, and, of course, I am really into movements. The Black Panther Party is really at the core of, like, anything I've tried to build. Angela Davis was cool, but now she has completely sold out into the establishment and is fragmentalizing the movement. I also read Adolph Reed Jr.'s *Class Notes*. It has a good analysis of why we need to centralize class over anything else." Shakur was convicted of the murder of a policy officer in a controversial trial, and I asked Tefa if she supports violence. She replied that she understands why some movements have been violent under specific conditions: "I come from a country that has misused nonviolence as a means of preventing workers to protect their lands and families."[28]

Aina Lakha is twenty-seven, has Southeast Asian ancestry, and grew up Seattle in a Shia Imami Muslim community. They are currently deputy chief of staff for Phara Souffrant Forrest.

Aina's history of activism began in high school around issues such as Islamophobia and the death penalty. Aina has explored a range of radical political and spiritual interests. At one point, they were involved with the spiritual organization Oneness University. In the early 2010s, Aina moved to New York City for college, studying comparative literature and society at Columbia. Aina continued exploring social movements and remembers a Black Lives Matter protest in 2014 triggering political interest. This experience motivated them to join the college-based Trotskyite International Socialist Organization (ISO), which was kind of a sectarian discussion club that dissolved in 2019. Aina has read extensively about the Russian Revolution and Marxist philosophy, particularly Lenin. They had a personal crisis in 2016 when they were transitioning and moved back to Seattle for a while.

Upon returning to New York in 2017, Aina participated in transgender activism and found that the situation on the Left had changed. ISO was in deep crisis, with dysfunctional leaders and members moving to DSA because they felt compelled by its electoral work and the Sanders campaign. There was strong skepticism in ISO about electoral work and Sanders because he brought the Left into the Democratic Party. Aina found themselves in a reformist, electoral democratic wing and was looking at the Julia Salazar campaign, thinking, "This is exactly what the Left should be doing." Around this time, Aina began dating a person on the Central Brooklyn Organizing Committee. Then, in early 2019, Aina joined DSA and went to a Central Brooklyn branch meeting, where they were thrilled by the decision to support the Sanders 2020 campaign. A few months later, Aina was elected to citywide leadership. What did leadership look like at the time?

> LAKHA: There was a lot of room for creativity. It might not have been as open as in 2017, where everything was possible, but it felt very alive and vibrant. We could put something together and figure it out—people just doing shit. Some projects were good, and some were bad. Nobody was totally proven right about what to do or what not to do. We had Julia in Albany, and people were excited about electoral work, but it wasn't the main focus of everything. It was exciting and serious. We're thinking through, "What's our strategy?"
> AUTHOR: Do you feel that your experience in ISO prepared you for leadership in DSA?
> LAKHA: Totally, I'd been active on the Left for five years. I knew how to speak. I knew how to write articles—"here's what a strategy is." It was also a humbling time because I was trying to learn and figure out "What is DSA, and how does it work?"

Aina quickly became an important resource for the organization: She was elected to citywide leadership within months and managed two victorious electoral campaigns in less than a year (the Jabari Brisport and Phara Souffrant Forrest campaigns). Aina helped set up the Socialists in Office Committee and served on the Health Care Working Group Organizing Committee and National Electoral Committee. How has Aina approached her leadership role?

> AUTHOR: How would you describe your approach to leadership in DSA?
>
> LAKHA: I bring the ability to think, study, and synthesize things into a political perspective and strategy. I bring an understanding of what being a member of this organization means. I also bring a certain perspective on power. A lot of people on the Left are afraid of power. They're afraid of responsibility. They're very afraid of getting yelled at. They don't want to take decisions they feel are conservative. And I'm usually not afraid of those things and try to be like, "No, it's good to be responsible. It's good to have power. It's good to be forced into situations where someone is putting pressure on you to make a different decision because it means you have something." I take on the burden of being the object of other people's projections, their frustrations and desires. That's part of what leadership is in a group. So I think this combination of political sophistication and will to power, which is an ethos in politics, those are things I enjoy bringing. Both come from my Leninist training. It's the kind of understanding that organization is a weapon tool for class struggle or whatever—analysis of the concrete conjuncture and the concrete situation.

Aina's political perspective evolved after joining DSA. She highlights the 2016 Sanders campaign as transformative because it showed that there can be "a working-class alternative to the neoliberal Democratic Party in electoral politics." Aina also felt that Hillary Clinton's loss demonstrated the necessity of this working class to overthrow the fascist Right: "[The neoliberal Democratic Party] couldn't defeat Trump because they represent rich people, not working people, not the majority."

Aina has a sophisticated political identity, defined by her attachment to basic principles of classical social democracy and her experience as a transgender person of color in New York social movements. Whereas some leaders in NYC-DSA have a specific focus on individual issues, Aina is motivated by a broader vision of integrating socialism with the racial justice movement.

I'm a kind of classical democratic socialist. A lot of my political references are in the best victories of left-wing social democracy in Scandinavia. I also think that Eduard Bernstein's thinking is important because it situates our political project in relation to two things: the expansion of democracy and freedom. Those two central pivots of what the socialist strategy means open up a whole rich kind of political universe. So we're trying to democratize the state, make it more accessible to people. And we're trying to expand democracy into the economic sphere through the state and through institutions for workers, like unions. Democracy is an institutional process, one of constructing something as opposed to tearing something apart in a revolution. The freedom piece has to do with the moral discourse on freedom that comes from the Black struggle in the civil rights movement, Dr. King's moral vision of a better society. His democratic socialism flows from understanding that commitment and seeing the gains of what a better lifestyle could be had in places like Scandinavia. I think of our project as trying to create this positive long march to transform things through struggling for democracy and freedom.

Aina has a deep knowledge of cultural theory and the democratic socialist movement and its electoral aspirations. Aina approached my questions with a holistic perspective and attention to the complexity of situations. Her main focus is on the practical leadership work required to help the organization develop and reach its goals. She also argues that the practical experience of (embodied) organizing and canvassing gives leaders in NYC-DSA a sense of how constituents feel and think. "Most politicians are just pretending to know how the community feels," she remarked. Finally, Aina is not channeling her intellectual resources into the philosophical space of the caucuses. She has been a member of Bread & Roses but feels that caucus discussions have been ascribed too much importance and have stolen attention from the fundamental project of the organization. To understand this project more fully, chapter 8 examines the party dimension.[29]

8

The Party Dimension

This chapter expands the organizational analysis to the party dimension. It begins with NYC-DSA's overall goal of democratizing the political process in response to the situation in the Democratic Party. I then show how the organization's party identity and position in the party system have evolved from the initial motivations to organize electoral campaigns. The analysis uncovers the potential for clarifying the party dimension and developing the leadership approach. NYC-DSA's party identity is ambiguous, because of its hybrid and informal nature. NYC-DSA is essentially a movement organization with party elements, and it has a movement party approach to creating synergy between movement and institutional contexts. It develops synergies between movement pressure campaigns and electoral campaigns, and movement leaders and electeds collaborate across movement and institutional contexts with a shared understanding of the political project and the organizational culture. With regard to party identity, members understand NYC-DSA as an alternative to the Democratic Party. NYC-DSA is not part of the party organization. It uses the party's ballot, however, so in the formal terminology with voters, it is a faction of the party and not a party alternative.

Democratizing Political Campaigns

The blueprint for NYC-DSA's electoral campaigns crystallized about one year after the organization initiated electoral work in 2016. However, the people who developed the organization had experience and learned from a longer history of movement organizing. Tascha Van Auken helped institute the community organizing approach popularized by the 2008 Obama campaign. She cochaired the original Brooklyn Electoral Working Group, and she managed the first three campaigns, in the process training future campaign managers. I use Van Auken as a focal point in this chapter, but it should be clear that organizers in NYC-DSA work in teams. Van Auken credits her collaborators Sam Lewis, Michael Kinnucan, Renée Paradis, and Asher Ross.

Van Auken joined NYC-DSA in 2017 with plenty of experience as a grassroots electoral campaign organizer and growing frustration about the lack of grassroots democracy in the Democratic Party organization (see chapter 5). Her approach to electoral work is focused on creating meaningful grassroots participation, defined by organizational egalitarianism and efficiency, as well as by politically intentional relationships. Van Auken feels that dominant approaches to political organizing are characterized by superficial routines, hierarchical gestures, and a problematic reliance on external consultants.

Van Auken's recollection of her early experience shows that the electoral project coevolves with other organizational changes:

> AUTHOR: What are some organizational changes in NYC-DSA that you have observed?
>
> VAN AUKEN: Overall, there's been a pretty regular evolution of challenging ourselves, learning new tactics, adding amendments to the constitution, and changing processes of how we do things, how we elect people that makes more sense for the growth of the organization. When I joined in 2017, there was just one Brooklyn branch, for example. And now [in 2022] there's three [a fourth Brooklyn branch was created in 2024]. A strength of this chapter is just being able and nimble, changing structures based on the growth of the organization. When I first joined, there was a pretty decent amount of electoral skepticism, people who didn't want the chapter to engage in electoral politics. We've been able to reimagine how to approach electoral work in a way that's baked in with other priorities. None of it is perfect or easy, but there's a pretty good amount of buy-in on electoral work.

The desire for an alternative to a conventional party organization becomes clear when Van Auken describes her approach:

> AUTHOR: What were the lessons from the first races? They were city council races, no?

Van Auken: Yes, and we didn't win either of them, but we developed an electoral strategy from that and codified it.

Author: What was the strategy?

Van Auken: Well, the traditional model for an organization making endorsements would be like, we have all this debate about the candidate: Are they good? Are they worthy of our endorsement? Blah, blah, blah. Okay, we endorse. What do we do? Let's send our members to the candidate and go work for the campaign. Our members get asked to do whatever by the campaign. The candidate wins or loses. We have built zero power for ourselves through this campaign. We reset at zero. The next time somebody's running, we're back at zero. We haven't really built any institutional knowledge or relationships or anything through the campaign. And the candidate is not really connected to us. In fact, most organizations, they endorse, and that's it. It's like a stamp. But they're not sending people to the campaign. I know this from running campaigns. You're not getting tons of volunteers every week from people who endorsed you. Sometimes you're getting a couple, which is great. In DSA, we don't do paper endorsements. We build foundational structures and leadership in the campaign, train people how to do campaign things, how to build a universe of voters, how to write a log script, how to set up phone banking, how to navigate the endorsement process, how to do everything—press compliance. And when we're endorsing, we're making a commitment to do a lot of work on the campaign.

In the Democratic Party, a candidate typically has no idea how to run a campaign. Some friend of a friend who's connected politically or somebody in the organization might be like, "Oh, well, you should just hire this political consultant to set up your campaign. They'll do everything you need." And so the campaign gives all their money to this consultant, who then builds their campaign, usually not very well. Consultants generally don't have a very high level of competency because they have no real stake in it. Their goal is to make you as a candidate think that this stuff is really hard and that you need them to make high-level strategic decisions. The reality is, most strategic decisions in electoral campaigns are pretty straightforward and not that complicated. Anyway, you give them all your money. They "run things." And then by the end, they own all the institutional knowledge built around your campaign, all the relationships. If you win, your relationship is with a political consultant or whatever insider political folks that they've connected to you.

Our model gets rid of the consultants, gets rid of that insider baseball. Instead, candidates become attached to hundreds of people who worked on their campaign and who are building something together. That was essentially our strategy that we came out of 2017 with.[1]

Instead of concentrating the decision-making and expertise in a handful of consultants, NYC-DSA incentivizes members to learn skills and take on responsibilities in the development of electoral campaigns. The idea is that members form grassroots committees that provide a political vision for the candidate beyond what consultants normally do for candidates in the Democratic Party. Sam Lewis argues that the commitment to grassroots democracy is part of the explanation for NYC-DSA's longevity and its unique position in the political field. Without meaningful grassroots democracy, volunteers would not have participated and organized for years, Lewis argues. In this perspective, the organization's focus on an electoral project rather than a party organization can be understood as a focus on the civic sphere of democracy rather than on the institutional sphere. This supports Aina Lakha's vision of democratic socialism as a constructive process of building community, justice, and welfare (see chapter 7). The grassroots-owned electoral project also distinguishes NYC-DSA from the anti-Trump movement, in which the grassroots participated in the 2018 midterms but did not create their own electoral organization. Lewis recalls:

What made the Electoral Working Group special and successful and continue on when so many other three-hundred-person meetings that happened in January 2017 didn't go anywhere is that we, from the very beginning, were like, "If you come to this meaning, you have a voice in this strategy. We are going to decide as a group who to endorse ourselves and take ownership of the campaign. We're going to master every aspect of the campaign. There is no expert here who's going to tell you what to do. There's not a leadership body that's gonna feed you endorsements. Join DSA and have a voice in this process." People very quickly developed a real sense of ownership. I think that is very rare. I really think we, from the beginning, had a commitment to democracy, collective ownership, and resistance to expertise.... Tascha Van Auken was one of the formative people in creating that culture and one of the most important people in building DSA's electoral work because she was so good at creating a political space where people could come in and feel like they had a voice, feel like they had a role and develop new skills.[2]

An important aspect of the organization's approach to democracy is the explicit political intentionality in communication with the grassroots. This perspective is highlighted by Gustavo Gordillo, who was trained by the founding members of the Brooklyn Electoral Working Group:

> Tascha created this idea that in an electoral campaign, everyone has to be treated as an equal and understand the strategy and the political context in order to do the kind of work that we're asking them to do. In electoral races, political consultants normally have all the political knowledge, and it's in their professional interest to keep that knowledge to themselves. For DSA, it's the opposite. We're structurally incentivized to get more members because we're dues funded and want members to work hard for the campaigns that we take on. The only way to do that is to have people really believe in a strategy and the purpose of what we're doing. So I think everything is designed that way to create that sense of identity and commitment. And having a clear ideology also helps for that.[3]

The Movement Candidate

The grassroots principles register in the figure of the movement candidate. This figure is a variant of the "movement leader" characterized by political expertise, integrity, and the capacity for base building (see chapter 7). AOC describes NYC-DSA's candidates and politicians as humble and distinct from the "more socialist than thou" attitude of the far Left.[4]

Whereas political parties frame their campaigns hierarchically around individual celebrity, NYC-DSA centers on the issues and the collective action. Van Auken talked in detail about how she cared about the Obama and Sanders campaigns, not the candidates.

AUTHOR: I assume that charisma and communication skills are important to a political candidate?

VAN AUKEN: It's great if someone is a great public speaker and charismatic, but I don't think it's our requirement. The word that I would use for it is "movement candidate"—somebody whose language is baked into being part of the movement. It's not about them. It's about a bigger movement and bringing people in, beyond their election and beyond them being in office. Over the last five years of endorsement processes, our members have become very good at hearing the right language from potential candidates.[5]

The movement candidate in NYC-DSA is also characterized by a positive movement psychology. The positivity is fueled by collective effervescence and the experience of shared affective and moral commitments, including belonging, solidarity, pride, and compassion. Positive psychology is central to the Ganz approach to movement leadership and Lakha's ideal of democracy as a constructive process. Leaders expected political candidates to channel these emotions and values into the campaign. They frequently used the phrase "nice to be around" when they described the desired qualities in a candidate. A campaign manager said that this is essential to building relationships and has to be in place from the start. AOC has described positivity as "an organizing tool" for NYC-DSA candidates.[6]

The positivity extends to the candidates' communication. They aspire to AOC's example of clear arguments, concrete solutions, and challenging opponents with courage but not bullying, ultimately seeking to demonstrate high moral standards.

The lack of grassroots democracy in the Democratic Party is interpreted as a moral problem. This is illustrated by Van Auken's experience in the Democratic clubs in Brooklyn. She spoke at some club meetings when she promoted "Team Bernie NY." In one meeting, a long discussion among members was interrupted when a group of electeds came in five minutes before the end of the meeting. The members responded with excitement about their appearance: "Oh, here's Assemblyperson so-and-so." The electeds did not participate in the discussion and did not listen. "They were just there to vote for Hillary."[7] I recognize this culture from my participation in a few club meetings. The organizers adopted a formal approach that mimicked institutional procedures unnecessary for such a small gathering of local members. Bureaucratic routines and formal presentations took precedence over communication about political substance with participants.

The Development of a Party Organization and Identity
A Structure to Sustain the 2016 Moment

NYC-DSA did not have a developed party conception when it began organizing elections. It entered institutional politics with a proposal for "electoral work," not a party. The grassroots did not have strong ties with institutional politics and were in the process of overcoming their alienation from it.

The proposal to initiate electoral work was a somewhat spontaneous response to national events, although it was informed by electoral organizing experience. The proposal primarily seeks to sustain the moment in the democratic socialist movement. It does not use the word "party." The motivational part mentions two Democratic primaries and thus implicitly suggests the building of an independent faction in the Democratic Party. This is one of the most influential documents in the development of the new NYC-DSA, so allow me to quote at length:

> Proposal for Brooklyn Branch DSA Electoral Committee
> Passed by Brooklyn branch membership 12/23/16
>
> Introduction:
> The success of Bernie Sanders's campaign for U.S. President signals that there is a passionate desire among millions of US voters for a new type of politics—democratic socialism!
> Bernie himself won his first elective office in 1980, when he was elected Mayor of Burlington Vermont as an independent socialist. While radical local campaigns have been few and far between in the US, recent years have shown a few bright spots. . . .
> Winning socialist electoral campaigns can be an important part of building the democratic socialist movement—we can use our victories to

challenge the billionaire class directly, build the confidence of working-class voters, and open up new opportunities for radical politics.

> Proposal:
> The formation of an electoral committee of the Brooklyn branch of NYC DSA to work on:
> - Recruiting socialist candidates to run in local elections
> - Developing and implementing a base-building plan for DSA so that tens of thousands of voters can be mapped and mobilized to support our candidates
> - Mapping districts and politicians to analyze where socialist candidates should run
> - Learning and teaching organizing skills like canvassing and campaign planning
> - Aiding the NYC chapter in developing questionnaires and interviewing candidates
> - Building the electoral wing of a grassroots democratic socialist movement![8]

The Central Brooklyn Electoral Working Group has largely achieved these initial goals, and electoral work has transformed the organization. Electoral work has helped sustain the movement, with electoral campaigns serving as a mobilizing structure, pulling in new members and organizing them into collective action. Additionally, electoral work has created a small base of DSA-backed electeds in government and brought the organization into the networks and cycles of institutional politics. It has entered the institutional environment of the political field. One might argue that the party dimension evolved from the electoral campaigns, with the movement organization gradually implementing party elements. The goal was never to explicitly create a party but to use electoral campaigns to boost the democratic socialist movement, accepting partial but not full institutionalization. Electoral working groups were created in neighborhood branches, and they collaborated with leadership committees and thematic working group committees. Organizational continuity in electoral work and attachment to the informal party identity developed through the reproduction of the approach to organizing campaigns.

In Search of a Place in the Party System

The arrangement secured a high level of autonomy for the grassroots and required no formal party organization. The movement simply used the Democratic ballot. The Dirty Break model was therefore a practical solution from a movement perspective. It became DSA's formal national electoral orientation at the 2019 national convention; but it has not stopped conversations about the organization's place in the party system: Should DSA be a faction in the Democratic Party or become a third party? This discussion is evolving, and my understanding is that NYC-DSA's position is hybrid. I see it as a movement-based organization with party elements.

The discussion is primarily happening in NYC-DSA. The national level of the organization has created barebone guidelines, mostly adopting principles from NYC-DSA. It also channels resources into electoral campaigns in NYC-DSA. The Central Brooklyn Electoral Working Group has had a strong representation in the National Electoral Committee. National DSA passed a resolution to participate in electoral work in 2017, deciding that candidates should target centrist incumbents without challengers, have a realistic chance of winning, be carefully vetted before getting national endorsements, be held accountable, and advance the movement.[9]

New York socialists initiated the discussion of party development in the summer of 2016, when Sanders lost the primary. Movement insiders asked how electoral work could sustain the momentum, but there was no easy solution in sight. Neal Meyer argued that the third-party model was unrealistic, as illustrated by the history of the Green Party. This party was lacking the resources to be a party and had run several presidential campaigns without winning over more than a tiny number of Democratic voters. Meyer also observed that people in DSA no longer had faith in Harrington's realignment model because democratic socialists entering the organization as individuals had not been able to create change.[10]

A more comprehensive and influential contribution was Seth Ackerman's article "A Blueprint for a New Party," published in August 2016 in *Jacobin*. It explores a place for the democratic socialist movement in the national political system. Ackerman first acknowledges the various obstacles facing insurgent political actors in the United States. He analyzes US election law to find that a new party cannot exist on equal terms with either of the two major parties. The law is designed to protect the elites and provides the major parties with measures to repress any challengers. Third parties are also faced with the onerous task of continually maintaining ballot status. Ackerman also uses the progressive Working Families Party (WFP; 1998–present) in New York as an example that a third party can have considerable local influence but will remain chained to the interests of the Democratic Party. WFP was initially a leftist alternative to the Democratic Party but became more centrist over time, and NYC-DSA filled the void. WFP endorsed Joe Crowley over AOC and Elizabeth Warren over Bernie Sanders.[11]

In Ackerman's view, the most realistic solution for broader political change is a transformation of the Democratic Party, but his suggestions show that there is not a quick and simple solution. Ackerman claims that popular democracy must involve parties organized by grassroots outside the political system. He suggests the creation of a national organization registered as a "party committee," in which candidates can run on independent or Democrat ballots, depending on the local situation, but with a single national label and political platform. This model calls for strengthening DSA's national level. A significant aspect escapes Ackerman's attention: voter attachment to party identity. Many longtime Democratic voters are reluctant to vote for a third party.[12]

The likelihood of transforming the Democratic Party seems further out of reach today in one respect. The national party leadership is fiercely maintaining its course. On the other hand, there are growing frustrations about its position, more inequality, and more turbulence in the political system. Those developments create conditions for major structural change. The question is increasingly whether such change will happen democratically or whether the country will regress further into oligarchic dictatorship and violence. The socialist movement was weakened at the national level by the pandemic and the Biden administration, and it suffered a major blow in 2024 when Jamaal Bowman and Cori Bush lost their congressional primaries because of their courage to stand up against genocide.

Challenges to Developing the Party Identity

The development of a party identity was complicated by changes in the national political environment that had inspired the insurgency. Sanders's loss in 2016 made it more difficult and less appealing to create a socialist faction within the Democratic Party. If he had won, the party could have become more aligned with the socialist movement and its grassroots networks around the country. Trump's election created further skepticism about national politics, and the shock prompted a plethora of disconnected efforts on the Left, some of them with the short-term perspective of electing young progressive Democrats in the next election. That was the goal of Justice Democrats and Brand New Congress, which were essentially political action committees.

Internally in NYC-DSA, the development of party identity is constrained by the movement-based model of the organization's macrolevel and by internal disagreements on this issue. The debates about the party identity and the relationship to the Democratic Party have evolved. Growing frustrations with the party's leadership in New York State have created further distance and led to calls for creating a new party. In practice, however, NYC-DSA works in coalition with progressives in the party, and this direction is advanced by the influential Socialist Majority caucus and the Ecosocialist Working Group.

While NYC-DSA is far from developing into a new party, participation in institutional politics creates expectations for a party identity. In the early 2020s, a structured effort was made to create a more distinct party identity and a more centralized party organization. Members of the Bread & Roses, Emerge, and Marxist Unity Group caucuses proposed a "1-2-3-4 Plan" at the 2022 NYC-DSA convention. The resolution proposes (1) the creation of coordinated and centralized communications and design teams, (2) that candidates should explicitly and publicly identify as democratic socialists, (3) greater alignment of candidate platforms, with candidates running as one slate on some of the same core issues, and (4) that candidates should commit to a set of rules that ensures loyalty and collaboration with the party organization, including the commitment to join the Socialists in Office Committee.[13]

A more aligned electoral organization and a clearer party identity would strengthen the organization's efforts in the electoral arena. It is necessary for creating stronger and more coherent political platforms for the campaigns and for strengthening communication with voters. The campaign materials and the training of canvassers do not communicate about party identity. The campaign communications do not frame the candidate around DSA, democratic socialism, or the Democratic Party. Instead, they focus on concrete issues of the campaign platform and the alternative to the incumbent. They personalize this messaging through a brief biography of the candidate. The core symbolic structure of electoral politics is not really addressed, and this lack of transparency is inconsistent with the organization's ideals of democratic accountability. The individual canvasser and voter cannot solve this problem.

There was broad support for the general ambition of the 1-2-3-4 resolution, but it failed to pass because of some of the specifics. I agree with the astute Chris Maisano that the resolution would have helped create more clarity, even though the relationship with the Democratic Party would still not be fully resolved. Maisano commented after the convention, "When we go and knock on doors, one of the most common questions we get from voters is whether our candidate is a Democrat. The resolution would have had our candidates attempt to pull off a rather difficult feat—win the support of the most loyal Democratic voters while simultaneously distancing ourselves from the Democrats *tout court*."[14] Maisano added that while the party debate continues to be disorienting at times, the core element of the Dirty Break model, namely, using the Democratic ballot and not creating a party, remains intact. Other core elements of the model have not been realized. Maisano continued, "While the dirty break has been the organization's formally expressed electoral strategy, our de facto strategy has been what [David Duhalde] cheekily calls 'dirty stay'—an approach that aims to use Democratic Party ballot lines without seeking to transform the party's institutional structures nor making concrete steps to create a new party." Maisano revisits the conversation initiated by Meyer and Ackerman in 2016, only to conclude that "the only viable electoral strategy is to continue operating as part of an organized, left-wing faction in Democratic Party primary elections and in legislative bodies."[15]

Another internal cultural challenge to the clarification of the party dimension is the drastic change and diversification of the membership in the new NYC-DSA. Most of the new members had little prior experience in the political field. They were not coming from political parties and were skeptical of party organizations. The low barriers to entry made it easy for people on the far Left to enter the organization, even though they were not aligned with its program. These so-called entryists would then try to push the organization further to the left and problematize its fundamental values, including the electoral project, making the old far-Left claim that participation in elections would only corrupt the movement because the state is a mere tool for the capitalist ruling class. The transformation of the membership was a dramatic and alienating experience for longtime members. David Duhalde, for instance, describes the new DSA as "an umbrella organization" and "a coalition of factions." There is some truth to this, although the situation is experienced differently by those who are participating in campaigns and experiencing social and political cohesion there. Some veteran members left in protest after the organization organized demonstrations against the genocide in Palestine.[16]

As a canvasser, I learned that the organization positioned itself implicitly in the realm of the Democratic Party by only targeting Democratic voters. Each canvass in the Alexis campaign began with canvassers downloading a list of Democratic voters in a small part of the district, typically a few blocks. The MiniVAN app guided us to the individual voter and helped track the results for later use in the campaign. We would begin every conversation at the door by confirming the identity of the voter. This level of filtering surprised me coming from Europe, where any citizen can vote for any party without voter registration. At first, I thought that the restriction to registered Democrats was a prioritization of resources, but I learned that it is a function of state law. New York City has "closed" Democratic primaries, so candidates registered to another party or no party cannot sign petitions or vote legally in the primary election. The campaign field guide offered the following template for an opening statement: "Hi! My name is ___. I'm volunteering for David Alexis, a candidate for state senate." I felt that it was more helpful and honest to say that I was volunteering for "a Democratic candidate in the state senate race" because the candidate was on the Democratic ballot. This also immediately helped answer the question that I would otherwise get: "Is the candidate a Democrat?"

In my experience, most voters have a remote relation to politics and primarily rely on information from the media. They easily get confused when the conversation extends beyond the two major parties. To alter this structure would require a considerable media campaign, including advertising in the controlled environment of broadcast television.[17]

When a constituent opened the door, they would often be skeptical and complain about political culture. I let them vent for a few minutes and found them to be open to dialogue after that, especially if they felt heard and were asked meaningful questions. My goal in this situation was usually nothing more than getting them excited about a primary challenger, motivated by concrete perspectives that indirectly represented democratic socialism. I learned that my canvassing colleagues also tended to talk about issues rather than ideology or party identity. I only brought up socialism when some level of mutual understanding had developed. This happened in about one-third of the conversations. I also learned that experienced and native-speaking canvassers were able to get further than I was. This is important to the climate dimension of the campaign, which is the topic of chapter 9.

9

The Road to Green New Deal Legislation

STATE OF NEW YORK
2023–2024 Regular Sessions
   IN SENATE February 3, 2023
   New York State Build Public Renewables Act
   [Adding fourteen new subdivisions to Section 1005 of the public authorities law]
   32. (a) Notwithstanding any other provision of law, [New York State] shall, on or after January first, two thousand thirty-one, only generate renewable energy and shall only purchase, acquire, plan, design, engineer, finance, and construct generation and transmission facilities for the purpose of generating, storing, distributing and transmitting renewable energy.
   (b) [New York State] shall prioritize funding, siting, building, and owning renewable energy projects which: (i) actively benefit disadvantaged communities as defined by the climate justice working group; (ii) minimize harm to wildlife, ecosystems, public health, and public safety; (iii) do not violate Indigenous rights or sovereignty.
   36. All new renewable energy projects subject to this section shall be considered public work, subject to articles eight and nine of the labor law and shall utilize a project labor agreement.
   37. [New York State], in consultation with labor organizations, shall develop a comprehensive plan to transition, train, or retrain employees that are impacted by the New York State Build Public Renewables Act, and shall establish and contribute to a just transition fund that shall make funding available for worker transition and retraining.[1]

The New York State Build Public Renewables Act (BPRA), quoted here, is a significant event in the history of climate legislation in the United States. It initiates a government-led transformation of the state's energy infrastructure. The provisions of the act are modest, in that they do not prompt a massive transformation, but they are nonetheless significant: The BPRA directs the state-owned New York Power Authority (NYPA) to build enough renewables to meet the state's 2019 landmark targets for reducing carbon emissions—the most ambitious targets in the country. Also, this is the first law in the country to introduce democratic socialist Green New Deal (GND) principles of labor rights, climate justice for Indigenous and vulnerable communities, and antitrust principles in the interest of lower-income communities. The New York State government passed another climate bill in 2023, the All-Electric Building Act, banning fossil fuels from most new buildings from 2029 onward, but that act regulates an individual element in the energy system, not its political outlook as a whole. The same can be said of the Climate Superfund Act of 2024, which also does not fund decarbonization. The Superfund Act mandates that fossil fuel companies pay for resilience efforts. It became an international media story, because it is easy to grasp and resonates with widespread public resentment against fossil fuel companies.[2]

BPRA is a movement victory and not the achievement of one organization alone; but NYC-DSA was the primary developer of this bill, and it developed and led the statewide coalition from its inception in 2019 to the bill's passage in 2023. NYC-DSA framed this legislative project within a broader campaign called "Public Power." The campaign combined elements of movement and party politics to develop long-term organizational capacity for the socialist movement in the climate crisis. The end goal is not just the individual bill or electoral victory.

The bill's passage in spring 2023 drew significant media interest and was celebrated by Bernie Sanders, AOC, and Naomi Klein. The underlying question in the dozens of interview articles published in the first couple of months was this: "How was it possible for the Ecosocialist Working Group (ESWG) to create and win something like BPRA?" The interviews provide valuable information about the process and its psychology; but the relationships and tactics in the overall process remain somewhat unclear. Knowledge of political organization is essential to providing this clarity. My account in this chapter focuses on the organization that made the process possible. I use and expand on movement party theory to explain how talented organizers mobilized beyond short-term protest. The analysis shows how they operated in a movement party organization and exploited opportunities in the political environment.³

My argument is twofold: (1) NYC-DSA created a new movement environment of climate politics, with dynamic capabilities to evolve and expand across movement and institutional contexts. (2) Leveraging this environment, NYC-DSA developed a proposal for legislation that climate organizations and progressive Democrats wanted but could not develop in their respective organizational contexts. NYC-DSA also developed and led a coalition with these other organizations, adapting to opposition and changes in the political environment. In the process, NYC-DSA's perspective expanded and transformed from the urban socialist movement world to a wider statewide public and the ecological transformation of the state.

The ESWG was an informal movement environment in the first years of its existence, but it combined movement and party elements from the outset. It operates as a movement setting of grassroots democratic exploration of perspectives on a "new" topic, in that climate was and remains underrepresented in political parties in the New York State government. The group has organized agitational pressure campaigns and is consciously independent of established climate NGOs, whose leaders are embedded in elite networks. At the same time, the ESWG's work is structured by the democratic socialist conversation about the GND at the national level.

The "Public Power" campaign was first launched in 2019 by the ESWG and evolved considerably, demonstrating movement party dynamism in NYC-DSA: It began as a pressure campaign for BPRA. Six months later, a coalition with external organizations was created to build the necessary statewide support. NYC-DSA kept growing the campaign for years through base building and by politicizing the energy system. When the bill was first defeated in the government's legislative session in 2021, the ESWG intensified both movement and institutional actions. The group threatened the most powerful politicians blocking the bill by exposing their fossil fuel donations and then primaried them. It electoralized "Public Power" into six ecosocialist electoral campaigns in 2022. The final push came from New York socialists in Congress, who wrote a public letter urging the governor to pass the bill. Throughout the process, the ESWG exploited opportunities in the political environment, above all building on the momentum of other climate legislation at both the federal and state levels. The congressional developments around the GND, Build Back Better, and the Inflation Reduction Act (IRA) played an important role, with the latter creating economic incentives for New York State to pass BPRA. Crucial to the success of combining movement and party elements, moreover, was the collaboration between movement leaders and electeds and the translation of the BPRA concept across movement and institutional contexts. The concept was translated from the movement into institutional language and then back into electoral campaigns and finally a statewide issue campaign.

The chapter begins by outlining the national history of the GND that created a context for NYC-DSA's climate work. The democratic socialist revival after the Great Recession created the ground for the seminal socialist definition of GND in 2019. During the height of the upheaval from Trump's election in 2016 to the midterms in 2018, the network of Alexandria Ocasio-Cortez's first campaign developed the GND agenda. She introduced a GND resolution to Congress in 2019 that was signed by seventy congressmembers, most of them from urban areas and deep-blue states. The resolution was administratively killed by the Democratic leadership, but it helped build momentum for the IRA of 2022 and the BPRA in New York in 2023. The GND remains central to the movement world of the climate Left.

The GND movement pioneered a socialist climate perspective in national mainstream politics and transformed the climate movement's role in the political process. It shifted focus from reactive to proactive activism, electoral politics, and political planning of the ecological transformation of society.

The remainder of the chapter explores how NYC-DSA developed BPRA. There are three major stages in this process: (1) The development of this new socialist movement environment of climate politics, 2017–2019; (2) the launch of the "Public Power" campaign for the BPRA and the development of the Public Power NY Coalition, 2019–2021; and (3) the expansion and escalation of the "Public Power" campaign through contentious protest, electoralization, and advocacy from socialists in Congress.

The New Era of the Socialist Green New Deal Movement

The term "Green New Deal" gained traction after Thomas Friedman introduced it in an opinion piece in *The New York Times* in 2007. Friedman's vision was essentially to create a greener economy through a stimulus package for the private sector. This is a defining aspect of the neoliberal climate policies that evolved from Obama to Biden and beyond to the European Union's "European Green New Deal" of 2019. These policies do not call for more just conditions for vulnerable communities and better labor conditions, even though they use the term "New Deal." The original New Deal reforms of 1933–1939 introduced legislation to recover from the Great Depression and create a social safety net, including guaranteed pension, unemployment insurance, the right to unionize, and a forty-hour work week. The welfare dimension has been stripped from dominant definitions of the term "New Deal" in neoliberal capitalism.[4]

Friedman's piece triggered interest among international leftist networks. These networks were, and still are, concentrated in the metropolis, with its educational institutions, think tanks, and international governmental organizations. A progressive London-based network called the Tax Justice Network developed a GND plan. The network included a former Greenpeace staffer, a few economists, *The Guardian*'s economic editor, and the Green Party's member of Parliament. Its report, published in 2008, adopted a focus on monetary policy and how financial globalization impacts the Earth's life-support systems. That same year, the United Nations proposed a Global Green New Deal, which included demands for wealth distribution in response to global fuel, food, and finance crises.[5]

The Green Party Pioneered the Idea

The pioneers of the climate Left in the US have worked for decades to connect the term "Green New Deal" with the New Deal reforms of the 1930s. One of them is Howie Hawkins of the Green Party. He had long been active in the socialist environmental movement that introduced demands for public investment in clean energy for job creation, economic justice, and environmental protection in the 1970s. Hawkins championed the incorporation of a Global Green Deal into the Green Party's platform in 2000. It was a demilitarization plan, a plan for global peace and fulfillment of basic human needs such as adequate food and clean water. Hawkins adopted a GND slogan in his campaign for governor of New York in 2010, and Jill Stein made it central to her presidential campaign in 2012. Green Party candidates thus brought the term into major electoral races in the early 2010s, but the party was not really part of the national GND movement that emerged a few years later. The young socialist leaders of this movement were empowered by the new socialist movement in the big cities and had a stronger institutional platform as Democrats in Congress, working closely with powerful progressive politicians. The Green Party also lacked the national prominence and talent of leaders like Sanders and AOC.

ESWG members knew that the Green Party pioneered the term "Green New Deal," but they were not inspired by the Green Party. One of the architects of BPRA told me that the Green Party is "totally irrelevant" because it does not organize and have a local presence, "besides, like, five old people who get together in a room to call themselves the Green Party. At any big GND event or 'Public Power' event, there'd be some old person from the Green Party, usually the same person, who would show up and were like, 'AOC stole the GND from us!' But you guys didn't do anything with it, so who cares?"[6] While I understand this perspective, I find it striking that core ideas in today's climate Left originate in the twentieth century. It tells us how long the fight for survival has been under way and why political organization is essential to winning this fight.[7]

Ocasio-Cortez in the Political Environment of the Trump Years

The political environment changed after the Green Party first introduced the term "Green New Deal" in the early 2010s. A few years later, climate change had become much more recognized as a matter of public concern.[8]

AOC became the central figure in the national conversation about the GND. She was the driving force behind the congressional resolution and is still the most vocal congressperson on climate. After proposing the GND, she played a key role in House Oversight Committee hearings targeting the oil companies, building on the legacy of other environmentally engaged Democrats, such as Timothy Wirth, Al Gore, and Bernie Sanders. AOC has brought critical journalists, scientists, and industry insiders to these hearings.[9]

AOC was moved into politics through environmental activism. One month after Trump's election, at age twenty-six, she took a road trip with two friends to the Midwest to get a "first-person idea of what was going on" in the country. They visited Flint, Michigan, and talked to people about the water crisis there. Then they went to Standing Rock Sioux Reservation, where the national #NoDAPL movement protests had been building for months. The federal government had approved a major pipeline construction plan in July that year, which had raised fears about the water supply for the Sioux tribe and potential oil spills. From October to December, the camp grew to ten thousand people protesting not just against the plan but also against the power of oil corporations and the government's treatment of Indigenous peoples. The measures from law enforcement were brutal. AOC found Standing Rock to be taken over by "militarized corporations," unchecked by political powers. She lived in the camp for weeks. At a New York town hall on climate change led by Sanders after her election in June 2018, she said, "I first started thinking about running for Congress, actually, at Standing Rock in North Dakota and South Dakota. It was really from that crucible of activism where I saw people putting their lives on the line, and Native peoples putting everything they had on the line, not just for themselves but for the entire water supply for the Midwest United States."[10]

AOC met activists at Standing Rock with whom she would eventually collaborate in subsequent years, including Sunrise leader Varshini Prakash. A few weeks after her return to New York, Trump won the election and expedited the pipeline construction, but this only seems to have hardened AOC. She decided to run for office and make climate a key component of her campaign. Her first action in Congress was to join the protest of two hundred Sunrise activists in November at Nancy Pelosi's office, demanding support for the GND.[11]

At home in New York City, AOC developed relationships with the city's burgeoning movements, and she became a catalyst for the ambition, talent, and grassroots movement enthusiasm that drive NYC-DSA. She became a member of NYC-DSA, which co-organized the field operation of her first campaign, and she worked with several leftist political organizations and local community organizations. For instance, she worked with New York Communities for Change on the campaign that prevented Amazon from opening a second headquarters in Queens in 2018. When she developed the GND resolution, she also talked to climate justice organizations in the city.[12]

AOC's first campaign for Congress and the GND resolution were not developed by local organizations in New York City, however. They were developed by two influential political action committees (PACs). AOC worked with Brand New Congress (2016–2023) and Justice Democrats (2017–present), both of which were founded by young people who had worked for the Sanders campaign and aimed to recruit and run young candidates across the country on the same fundamental platform. The two PACs gave priority to ethnic-racial diversity and the GND. They helped get a handful of socialists elected to Congress in 2018, known as the "Squad."

Justice Democrats was responsible for the platform in AOC's first campaign. It made a GND one of the four pillars of its platform and created the think tank New Consensus specifically to develop the congressional GND resolution. Saikat Chakrabarti founded Justice Democrats and New Consensus and served as AOC's chief of staff while coauthoring the GND resolution. His ambition was to change the policy discourse on climate. He wanted to get past the "climate versus jobs" dichotomy that framed the ecological transformation as an economic problem. Other people in New Consensus emphasized racial justice, including Rhiana Gunn-Wright, who has argued that historical injustices continue to perpetuate if they are not addressed. Overall, New Consensus was a small group of elite-educated democratic socialists with the ambition and skills to develop a historically deeper and culturally more diverse vision for the GND.[13]

The Green New Deal Resolution of February 2019

The GND resolution is an important political document and provides deeper insight into the emerging discourse of US ecological socialism. It also shaped the terminology of the ESWG. Readers familiar with the details of the resolution can skip to the next section.

The first thing to note about this resolution is that it is a proposal to develop legislation.[14] It can be viewed as a motivated call for action, a communication from a progressive minority to act on the climate crisis on moral grounds. The resolution was killed by the Democratic caucus leadership, which forwarded it to eleven committees for an indefinite period. The resolution eventually expired. The same thing happened when it was reintroduced in 2021. Its central principles of social justice were not adopted into the IRA of 2022. Hence, the GND resolution has not led to legislation but remains a powerful framework for climate thinking on the Left. I consider it a significant achievement of humanity, a vision for socially just climate survival in the Anthropocene, on par with the Universal Declaration of Human Rights of 1948 and the Civil Rights Acts of 1964–1965.

The resolution has a broad and ambitious outlook. It recognizes the extraordinary complexity and scope of the climate crisis. This is possible in the genre of a resolution, compared to the need for more concrete decisions in legislation. The narrative is that society should look critically into its past and prepare for a century-long transformation because of an unfolding global ecological disaster. The resolution suggests the development of large-scale changes to industrial and social policy with implications for society as a whole for several decades. It acknowledges cross-committee jurisdictional implications.

The opening three-page preamble outlines the challenge. It seeks to establish factual evidence of the climate crisis and other crises in society, stating that these crises are related and cannot be solved separately. The resolution summarizes the findings of two large-scale scientific reports published in the fall of 2018, each of which is the result of work by hundreds of scientists and based on a wealth of scientific publications. One of the reports, *IPCC 2018*, was conceived in the wake of the 2015 Paris Agreement and focuses on the consequences of a temperature rise to 1.5°C above preindustrial levels and the measures necessary for staying below that temperature rise. The IPCC is the UN's chief scientific body for climate change assessment and adopts the UN's ethical agenda of global social and economic justice. The IPPC provides evidence that the consequences of climate change fall "disproportionally on the poor and vulnerable." The GND resolution incorporates global climate justice discourse from the UN. A key element of this discourse originates in the economist Amartya Sen's moral argument in the 1980s that wealth should not be measured in aggregate terms. A country can have a high GDP but many poor people, systematic social deprivation, and neglect of public facilities. This thinking is antithetical to neoliberalism. The first paragraph of the resolution concludes that temperature rise must be kept below 1.5°C, pointing to the catastrophic 2.0°C scenario described by the IPCC, which is becoming a reality in the mid-2020s.[15]

The remainder of the preamble details the resolution's political argument. The resolution suggests that the United States has a moral responsibility to take a leading role because of its carbon emissions—the highest in the world per capita—and its high technological capacity. It suggests, moreover, that environmental problems exacerbate existing societal problems, described as "systemic injustices" for "frontline and vulnerable communities."[16] These problems include declining life expectancy because of pollution and lack of access to adequate health care. Climate change is thus interpreted as a public health issue. A clear leftist perspective emerges in the statement that climate change is exacerbating the consequences of rising inequality and their articulation across racial and gender lines. We learn about four decades of wage stagnation, antilabor policies, and inadequate resources for public-sector workers. Meanwhile, the top 1 percent has become richer. The solution is to revive elements of the New Deal, while also redressing its injustices. The New Deal helped create a large middle class but excluded frontline and vulnerable communities.

The resolution itself is a brief outline of the goals, followed by a detailed ten-year mobilization plan. The central goal is "to achieve net-zero greenhouse gas emissions through a fair and just transition for all communities and workers."[17] This involves the creation of millions of good jobs and prosperity for all as well as investments in infrastructure and industry. The mobilization plan in sections 2–4 defines tasks for developing greener and more resilient infrastructure in transportation, energy, and industry, particularly manufacturing and agriculture. It also addresses the political organization.

The socialist dimension in this plan does not lie in the individual climate goals but in the emphasis on equality, justice, and workers. The resolution calls for "stopping current, preventing future, and repairing historic oppression." It suggests partial public ownership and government expansion in response to the growing need for public infrastructure, education, and community assistance. It demands that vulnerable communities and workers be consulted and co-lead the green transition. These population groups should be offered education to be able to participate fully and equally. Labor unions and worker cooperatives should be supported and involved. The resolution defines "good jobs" as "union jobs" that pay "a family-sustaining wage" and give workers the right to organize and bargain without the fear of coercion or harassment. The classic social democratic principle of care for families and the life cycle is evident in the demand for "adequate family and medical leave, paid vacations, and retirement security to all people of the United States." Universal health care and housing should be guaranteed.[18]

In conclusion, the GND resolution has strong elements of classic social democracy and racial justice. The contrast with the dominant neoliberal policies is further evident in the fact that the resolution does not focus on private-sector growth.[19]

A New Environment for Climate Politics in New York City (ESWG, 2017–2019)

A climate change working group was created in NYC-DSA in early 2017, just two months after Trump's election. So climate was a focus area in the "new NYC-DSA" from the start, but it did not define the organization publicly until years later. In the first two years of its existence, the group developed its internal organization and political agenda from the ground up and was a junior partner for established climate organizations. By comparison, the Central Brooklyn Electoral Working Group began running campaigns in 2017 and won its first-New York State Senate race in 2018.

Much of the EWSG's distinctive approach derives from the fact that it leveraged the potential of the major changes in the national political environment in 2016. The ESWG connected the socialist movement with the climate field. It attracted some of the most talented young activists and organizers from the emerging socialist movement and a few climate professionals who had been involved in the climate movement.

The ESWG was a distinct environment in NYC-DSA from the outset, created by the focus on climate, connections with the climate field, ambitious and talented members, and gender diversity. A person with organizational responsibilities in the group from 2017 to 2023 remembers that it became his "movement home" but that he was "a bit cloistered away from the bigger-picture dynamics in the chapter."[20] The ESWG organizers had little direct interaction with Tascha Van Auken and the Central Brooklyn Electoral Group until 2021, although it adopted NYC-DSA's general platform and approach to internal democracy and collective action. Geographically, the group was attached to Lower Manhattan and not as exclusively concentrated in Brooklyn as other parts of the organization have been. Several meetings in the early years were held in places such as the LGBT Center in the West Village. Several active members lived in Lower Manhattan, which was also a central area for the GND conversation in the city. The conversation was stimulated by people associated with New York University's Department of Sociology and Institute for Public Knowledge. The ESWG became a small and focused space, somewhat removed from the more mass-public spaces in NYC-DSA and the conversations between political factions.

The most active members at first included Jamie Munro, Jamie Tyberg, and Magenna Brink. They were instrumental in bringing in talented and diverse people and creating this focused space. Over time, the group developed into a semiprofessional entity, with the majority being very active and having organizational responsibilities. Members found the environment exhilarating and believed they could achieve unique results with this group. They would not have devoted years of their lives if the group had been dominated by casual participants.

The story of the ESWG also brings nuance to the organization of movement and party elements in NYC-DSA. The ESWG was not involved in organizing electoral campaigns until the final stage of the "Public Power" campaign. It spent the first two years developing a plan for climate action, while the Central Brooklyn Electoral Working

Group was running electoral campaigns. Eventually, the ESWG would become a powerful force in NYC-DSA, take over the operation of all electoral campaigns in 2022, and use these campaigns as a movement tactic to win BPRA.

A central objective of the ESWG was to build long-term capacity without primarily relying on institutional representation. The group has struggled to achieve this goal, and I shall later explain why and reflect on the challenge of creating organizational stability with limited institutional infrastructure in conclusion 1.

Whereas the Electoral Working Group primarily positioned itself in relation to the Democratic Party, the ESWG also positioned itself in the field of climate NGOs.

ESWG members mostly did not have expertise in the climate area, and few were experienced organizers. Longtime members Charlie Heller and Gustavo Gordillo have emphasized this aspect. Says Heller, "We didn't know anything when we started out. If we can come together and do this, anyone can."[21] I can understand why members say this, but their educational backgrounds and social capital are not ordinary; so I would qualify the statement by saying that many people can do something similar and that popular participation can be expanded further. Starting from a tabula rasa allowed the group to develop a new culture and vision from the ground up. The group was also not insular.

However, the group had a few members with considerable experience and connections in the climate field. It tapped into resources of its broader urban and media environment, illustrating a further instance of the compression of time and space in modernity, with young political actors being lifted out of local and traditional contexts and into broader flows empowered by technology. The ESWG is deeply influenced by climate conversations among national experts, of which there are many in New York City, and by climate coverage in international publications such as *The Guardian*. The group took particular inspiration from Naomi Klein's *This Changes Everything* and Jane McAlevey's writings. Their writings were recommended reading for everyone and constantly referenced, and the authors interacted with the group, especially Klein, who lived in New York from 2018 to 2021 and came to see the ESWG as an important part of the movement for a new form of democratic ecological socialism. Klein describes how DSA incorporates climate justice and connects "the dots between the economic depredations caused by decades of neoliberal ascendency and the ravaged state of our natural world."[22]

However, it took years before the ESWG gained a wider public presence. The early years were focused on developing strategy and political goals. The group also developed climate platforms for the successful DSA state senate races of Julia Salazar in 2018 and Jabari Brisport in 2020.

One of the members of the group's Organizing Committee from the beginning in 2017, Gustavo Gordillo, describes the group's early history as follows:

> GORDILLO: We were setting up ESWG in 2017. They had never had a campaign. In other parts of the chapter, there would often be one person who had been organizing before 2016 and had the experience and the knowledge. Everyone looked to them as authorities in their field. A lot of the trajectory of the chapter depended on them. But ESWG didn't really have anybody like that. I started going to the Housing Working Group a lot because they had more experienced organizers. A big mentor for me was Cea Weaver from the Housing Working Group. A lot of our principles came from her. Early on, I would ask her questions all the time, like how she set up the Housing Justice for All Coalition and how they developed the Housing Working Group in DSA. We didn't have someone who knew the climate movement inside out for ten years. She said, "Yes, but that means that you can build something stronger than the Housing Working Group."
>
> We put a lot of importance on democratizing strategy within the group and developing a deep bench of leaders. We had lots of strategy retreats, sessions, and conferences, both in New York City and also nationally later.
>
> AUTHOR: What were some of the decisions on strategy resulting from this work?
>
> GORDILLO: We understood that the city government is seriously constrained, and the state government in Albany is where the most important decisions happen. The state has a lot of power over the energy system. The city does not. We also knew that a lot of

the tactics that nonprofits were using would not really work. DSA members just wouldn't be interested in a lot of insider negotiations or even lobbying necessarily. For us in ESWG, the crudest way to put it is that we believe that "bullying works." We win more through negative pressure and conflict than through conciliation with politicians, and there are only a few ways that we can exercise our leverage. Primary elections are probably the most important tactic that we have in NYC-DSA. So we started to learn how to plan rallies and actions, things like that, which we do often to target politicians and pave the way to run our own candidates.

Many other organizations, many nonprofits, have this understanding that when you work in coalitions, you're supposed to defer to some other authority. We rejected that. We thought that it was more important for us to develop our own strategy and our own demands. That was how we developed our own direction.[23]

My research suggests that there were a few experienced climate professionals and activists in the group and they played a key role in developing a coalition with grassroots-based climate NGOs, not elite-funded legacy NGOs, and lobbied in state government.

Shay O'Reilly and Patrick Robbins had worked in the climate field for years and joined the ESWG shortly after its founding in 2017. They were already well connected in the New York climate field. Patrick, for instance, graduated in climate and society from Columbia University in 2011, wrote several articles for Naomi Klein's blog *The Leap* from 2014 to 2017, and is longtime friends with Rajiv Sicora, the research director for Klein's *This Changes Everything* and, later, a staffer for Jamaal Bowman. Since 2017, Patrick has been the coordinator of the New York Energy Democracy Alliance, made up of twenty-seven community organizations "working together to advance a just and participatory transition to a resilient, localized, and democratically controlled clean energy economy in New York state."[24] He helped make this organization a member of the BPRA coalition. ESWG members employed in other climate organizations have similarly helped make those organizations members of this coalition. In 2021, Patrick and fellow ESWG members Thea Riofrancos and Daniel Aldana Cohen were fellows in the policy think tank Climate + Community Project and cowrote its report supporting BPRA. At a party at Kate Aronoff's house the following year, Patrick met Jamaal Bowman and encouraged him to pressure the governor to pass BPRA, which Bowman did together with eight other House Democrats from the city. The congresspeople wrote a letter to the governor urging her to pass BPRA. The letter got attention in *The New York Times*, prompting the first coverage of BPRA there. The ESWG had not been able to get this coverage on its own. The *Times* article helped amplify the pressure in the last month before the bill was passed by calling out the governor for not agreeing "to mandates that the power authority build to meet climate benchmarks." The article incorporated the campaign's appeal to New York's role as a national climate leader, claiming that the state's 2019 climate act was a partial inspiration for the IRA, and the article quoted the letter: "When New York leads, the nation follows." The ESWG has since had a close relationship with Bowman's office.[25]

Shay and Patrick's key contribution to the ESWG is steering it toward the energy sector, proposing a relevant and feasible idea for a climate bill with exemplary potential, and managing the relationship with key institutional actors in the political process until the bill was passed in 2023. Why did Shay and Patrick leave climate protest organizations and channel their resources into NYC-DSA and the ESWG more specifically?

>ROBBINS: I joined DSA in 2017 after being a part of the movement against fossil fuel infrastructure for a good part of my life. I got my start in 2013 fighting against a fracked gas pipeline, a methane pipeline that they were building into New York's West Village, where my mom was from. So I had been a part of the climate movement and worked professionally for climate-oriented nonprofits, but [the ESWG in 2017] was the first time I'd seen such a surge of people concerned about the climate crisis with an anticapitalist analysis and a socialist analysis.
>
>AUTHOR: The first time?
>
>ROBBINS: Across continental North America, there were plenty of groups that had an anticapitalist analysis but didn't necessarily have a strategy that reflected it. I say that with a lot of love and respect for individual people. There were a lot of talented and strategic think-

ers, but this was the first time that I'd seen that, at the scale that I saw it and in a way that had a clear-eyed understanding of what it was going to take to build a base for ecosocialism in this country. It's not that there was a fully fleshed-out plan, but there was an orientation to strategy that I appreciated. A lot of people there were able to think in terms of building power. I was more aligned with anarchism when I joined, but I recognized that there was a lot of potential there and eventually became a socialist.

When the ESWG was trying to figure out what its priorities were going to be, I and a small number of people pitched the idea of focusing on the utility system as a crucial site of struggle, where it's possible to make a lot of gains. The utility system is ultimately regulated at the state level, is widely hated by many New Yorkers, and wields an enormous amount of power over whether or not we are building renewables and meeting our climate goals. We pitched that in February 2018, and that was essentially the beginning of the "Public Power" campaign.[26]

Shay has a different social background but similarly joined the ESWG because it was a new and powerful alternative to existing climate organizations, and he, too, became a democratic socialist in the process:

My background was in economic justice work. To me, environmentalism was for people who could afford to go to exotic destinations like Costa Rica and fall in love with the rainforest. I grew up in a family that couldn't afford to travel abroad. I developed a relationship with environmentalism when I was doing economic justice work and started talking to people who were like, "Oh, no, climate change is our biggest issue because we live in a trailer park that got flooded out during a superstorm, and we had to organize and fight to get any kind of compensation for that to recover and rebuild our lives. And we know that the next storm that comes through, we're still in the trailer park that's at the bottom of the river valley, and we're just going to get flooded out."

I started to realize that climate change is the most urgent issue facing humankind. And if we address it correctly, we can amend some of the historical injustices along the lines of class and race. We can make a better and more livable world for everybody, a more just world. I then worked at an academic center on religion and climate and got involved with an anti-pipeline direct-action campaign up in Westchester County, supporting people who are climbing into the pipe and chaining themselves there. I then got a job at the Sierra Club, which is this very institutional NGO. I was like, "Ah, it's a bunch of white bougie environmentalists, but it's a job. Got to pay rent." It was fine, but the fall of 2016 was an incredibly apocalyptic time, and my NGO world was supporting private companies in getting offshore wind contracts in New York State. I was like, "Wait a minute, we're going to build these offshore wind projects in ten years, and in the meantime, we're going to miss the window for keeping warming within 1.5 Celsius? We need to do everything that we can to not miss that window. We need drastic action, but we also need to build the social infrastructure to be resilient and responsive in a circumstance where we missed that window. We need to bring more things under social control, inspire radical change, and give ourselves the policy tools that open up possibilities in the future when we have better political tailwinds and the right people in office." That's why the ESWG was so exciting to Patrick and me. It brought something to the table that other groups would not. Many of the people who joined did not have a background in environmental advocacy, but they felt that the ESWG and not the local Sierra Club was going to propose action on the scale that was needed commensurate with the crisis.

There were people with organizing experience in the early days of the ESWG, and we built our experience and approach to power into the group. From the beginning, there was this understanding that all forms of power are fake, except for the number of votes so you can kick people out of office. Money, PR, communications—all fake! The only thing that really matters is if you have the votes to back it up, and you are making a credible threat every time you say to people with power that you're going to remove them from power unless they do what you want them to do. That was our orientation from the beginning. Increasingly, that approach has spread more in the New York climate space, that real hard-nosed orientation that you've got to organize, or you can't get anything done. Legacy environmental advocacy groups have a brand that means something but less than they think it means. They think that being able to put their little logo on a politician's fliers when they're running for reelection matters, but in the end, it doesn't matter. There's no way that any of the existing legacy environmental organizations would have developed BPRA. It's just not possible.

Patrick and I had worked with climate justice groups and knew that they had reservations about the GND framework because of the racial exclusion that was part of the original New Deal, but the GND changed the situation. It was very exciting to see that move from divestment as the most prominent framework of the climate movement in the United States—divestment occupied a lot of energy in the early 2010s and these pipeline campaigns. There was a lot of, like, "No!" The GND connected it to a positive vision.[27]

The lengthy quotes from Patrick and Shay clearly show that the ESWG evolved in the broader context of the climate field—the field of climate protest, advocacy, and policy—and that the ESWG gained significance as a distinct alternative within this field. At the same time, the ESWG synthesized ideas and experiences from this field into a socialist movement environment. The ESWG did not start from nothing, but it created a new organizational and cultural context.

The Bill and the Coalition: A Turning Point (ESWG, 2019–2021)

This section details how the ESWG developed its political plan and the organizational capacity to carry it out. The group's approach in the first two years resembles a textbook design process of a broad exploration of problems and solutions, followed by the selection of one solution and then concentrating on its development and implementation. In addition to this political plan, the ESWG developed leaders and organizational structures.

The decision to concentrate on a plan for transforming the state's energy sector was made in a large workshop in 2018. There was a broad discussion of the possible paths that the group could take. Some members suggested that the group could take action in solidarity with other groups. There were invited speakers from organizations such as XR and UPROSE. Others argued that the group should take on its own project in alignment with democratic socialism. One of them was Shay, who had been inspired to think about the New York Power Authority's potential in a conversation at work. The New York energy system is dominated by ConEdison, with NYPA as a small actor that only provides electricity to public entities, not residential consumers. Therefore, NYPA had little public attention and had not been addressed in conversations about climate change. Shay pitched some of his fellow ESWG members on his idea: "Hey, guys, we can expand NYPA's ability to build and operate renewable energy. We have a public authority that basically just serves as a piggy bank for Governor Andrew Cuomo. Why is it doing that instead of building what we need? Let's expand NYPA's authority and make it the decarbonization engine for New York State."[28] Shay operated behind the scenes because of his public role with the Sierra Club. The idea was therefore presented at the 2018 meeting by Thomas Niles and Sarah Lyons. The idea initially included a slate of bills to take on different aspects of the electric system, but only BPRA survived. Developing BPRA was so demanding that the ESWG decided to deprioritize the other bills by the end of 2020.

The idea of transforming NYPA struck a chord with both socialists and the wider public. Classic socialist arguments for state ownership seemed relevant to the group's analysis of the state's energy sector, and many New Yorkers hated the energy monopoly ConEdison for its exploitative practices. Complaining about ConEd's rate hikes is folklore in the city, and the ESWG discovered that ConEd prioritized affluent neighborhoods during power outages, leaving lower-income neighborhoods in the dark during heat waves.

A proposal for an "Energy Rights" campaign in January 2019 set the goal of passing a bill in the 2020 legislative session to transform NYPA, and the name "Public Power" was adopted in a proposal in May 2019. The "Public Power" campaign began publicly politicizing rate hikes and power outages, as they were happening in Flatbush, the district of State Senator Kevin Parker.[29]

During 2019, the ESWG conceived "Public Power" as a broader movement vision of the state's ecological transformation, with BPRA as the centerpiece of the first phase. The vision includes political goals for sustainability and organizational goals for the political process, namely, the building of long-term movement capacity for responding to ongoing challenges in the climate crisis and ensuring involvement from below. This internal organizational narrative was clear to all my informants. It was also clear to all that there is a long way from BPRA to the end goal of "Public Power," which is the transition of the entire energy system into renewables, public ownership, and democratic governance.

While some group members organized the "Public Power" campaign, other members began developing a coalition around the bill. The coalition was part of the "Public Power" campaign strategy from the start, as seen in the May 2019 proposal. Robert Carroll, a progressive Democrat of Park Slope, was one of the few legislators who would shoulder the bill in the beginning, but legislators were not much involved until 2020, when NYC-DSA had more electeds in government and was preparing the bill for the 2021 legislative session. Community-based climate justice organizations were more involved in the early stages. Shay remembers,

> The concept had to be workshopped around with existing climate justice groups, especially because many of these groups had a lot of national sway, like UPROSE was part of the Climate Justice Alliance. Patrick and I had worked closely with these groups in our previous work. Their opinion of DSA more broadly was, "It's a bunch of white people with college degrees. They're not our people," which is not right, but it's also not wrong for them to think that. Patrick and I are white people with college degrees, but we had proven that we could show up and work with folks. We learned that these climate justice groups also had a history with NYPA. Peaker plants had been built in their communities in 2003 against their will, and they have been fighting NYPA ever since. The plants were supposed to be there for five years, but they're still operating more than twenty years later. In those conversations, some of the work was to build trust, and we needed to include stuff about environmental justice and retiring the peaker plants.[30]

The "Public Power" campaign was elevated to a citywide priority campaign in NYC-DSA in September 2019 with a proposal that embedded the campaign into "a broader NYC-DSA Green New Deal strategy," which includes areas such as housing, labor, transportation, and immigration. This expansion has not evolved far due to resource constraints, so "Public Power" has remained the central project. However, the proposal shows that the national GND helped move the climate agenda forward in NYC-DSA (see figure 9.1). The proposal was written by a

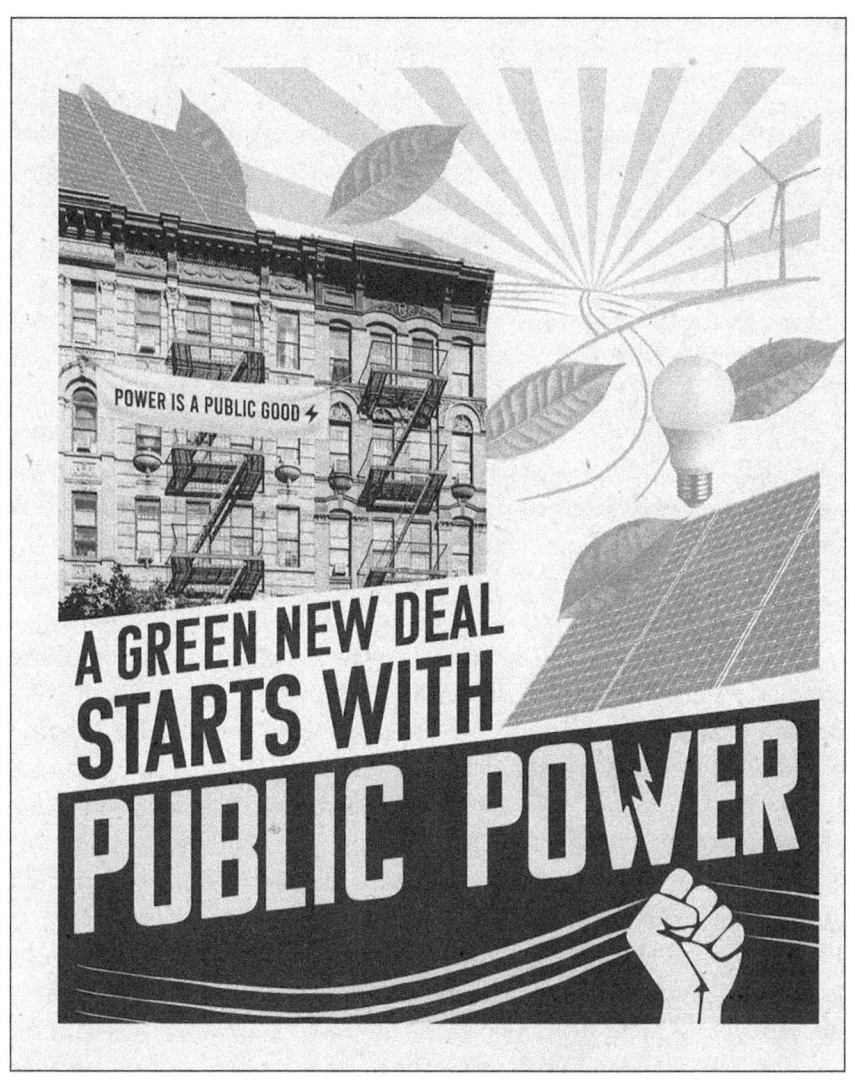

Figure 9.1. The NYC-DSA Ecosocialist Working group poster "A Green New Deal Starts with Public Power," by Meaghan Elyse Lueck and Lawrence Wang.

group of emerging ESWG leaders with vital roles in subsequent years, including Aaron Eisenberg, Andrea Guinn, Gustavo Gordillo, and Michael Paulson. The ESWG also jumped on another opportunity in the political environment, namely, the Climate Leadership and Community Protection Act (CLCPA) of July 2019. This act set goals for the state's greenhouse gas reductions, and the ESWG framed the BPRA as a path to this goal. This became the chief rationale among supportive legislators, who argued over and over that BPRA was necessary to reach the goals of the CLCPA. The ESWG also built on the coalition around CLCPA and included actors that Cuomo had pushed out for unrelated reasons, such as labor unions and the New York Families Party.[31]

The CLCPA stimulated the process in the ESWG. It was around this time that the "Public Power" campaign gained traction in the group and became clearly defined as an ecosocialist campaign. "That's when things really took off," remembers Patrick. There had been separate little working groups up until then, but after "Public Power" became the official priority of the group, a lot of the effort coalesced around that.

The final step in establishing the campaign was the creation of the statewide Public Power NY Coalition in November 2019. The ESWG needed partners across the state to win BPRA, and it helped define NYC-DSA as a serious actor in the climate field. The coalition includes twenty environmental community and advocacy organizations. Important partners include the New York Energy Democracy Alliance of community organizations and policy experts. The coalition was created at one of the Alliance's meetings. Patrick remembers, "I worked for the Energy Democracy Alliance and invited comrades from the ESWG to join the Alliance and come to some of our retreats. It was on a November retreat on a little farm in upstate New York in 2018 that the coalition was formed. I remember it clearly because a number of the member organizations that I worked with professionally—Metro Justice, Sane Energy Project, Alliance for a Green Economy—agreed to join a coalition around the idea of public power. Not all of my members joined, but a significant enough number of them joined that the coalition was formed." Other coalition members are the New York office of Food & Water Action, a national advocacy group; For the Many; the Long Island Progressive Coalition; and Sunrise. Shay, Patrick, and others communicated with labor unions to incorporate labor language, but the big unions were generally unwilling to work with the coalition until 2022, when it became clear that the BPRA might pass.[32]

This institutionalization of the Public Power NY Coalition in the climate field resulted in the development of the movement narrative into institutional genres. A matured rationale for BPRA can be found in the Climate + Community Project's 2021 report *A New Era of Public Power*, coauthored by three ESWG members. The report is not explicitly socialist and frames the agenda of socially just decarbonization in more general terms. It does not mention the bill but lends support to its core propositions through an evidence-based evaluation of the state's energy sector. The report thus translates the movement's vision for institutional actors, and it was used and referenced in the influential letter from nine US House Democrats to the governor in late March 2023, about one month before the bill was passed.[33]

The report begins by appealing to the unique potential: NYPA is the largest state-owned energy provider in the country and is uniquely positioned to carry out CLCPA mandates because of its ownership structure. NYPA is not governed by a focus on quarterly earnings and can reduce energy consumption, whereas private businesses are incentivized to grow consumption. The report further argues that the public ownership status can be developed into democratic governance and proposes a multistakeholder model with representation from labor and community organizations. The key proposition is that NYPA should be expanded to create more renewable energy and become a public option for residential consumers in the state, making them less dependent on for-profit energy service companies and "their predatory operations." These changes would strengthen the state's connection with citizens, lower bills for users, and revive NYPA's original purpose as an alternative to private utilities. The report thus uses antitrust policy rhetoric.[34]

Escalating and Electoralizing "Public Power" (ESWG, 2021–2023)

The first attempt to pass BPRA in 2021 was a transformative event, as one might expect. The years-long efforts preceding this event explain the group's strong reaction when powerful politicians tricked the group into failure. Kevin Parker told the coalition that he would support the bill if it could muster support from other legislators. This was foul play because it is not customary for legislators to support a bill unless the lead sponsor asks them to. The ESWG felt betrayed and initially responded somewhat spontaneously with a blockade outside the government's offices in New York City. The protestors held big poster boards with Venmo pages showing how much money powerful politicians in state government had received from fossil fuel corporations (figure 9.2). Photos were shared on social media and elicited strong reactions from the politicians. Patrick got a call from a chief of staff urging the group to take down the Venmo boards and describing them as "unfair." His reply? "Your boss did take this money, and it's not our job to cover up for them."[35]

The protest represented an escalation of the "Public Power" campaign's communications. The group began threatening its powerful opponents by exposing their moral failures and eventually also by primarying some of them. Legacy environmental organizations would not have done this because they are friends with these politicians and belong to the same elite networks. The communications team, led by Charlie Heller and Lawrence Wang, began doing more dramatic and agitative communications. In particular, it contrasted images of a green New York and a New York in flames, accentuated by the question, "Governor Hochul, do you want the state to build or burn?!" The bill was thus promoted through contentious movement communications in noninstitutional channels, including social media. The relationship between movement and institutional politics became more intense.

The pressure campaign intensified further when the bill was derailed a second time in 2022. Thousands of activists participated in the campaign during the peak months leading up to the budget negotiations in May. More than fifty thousand phone calls were made each week encouraging citizens to ask their representatives to support BPRA. The bill passed the senate and just needed the final step of being approved in the assembly. There, Speaker Carl Heastie prevented the vote at the last minute. He was silent for days but eventually made the dubious claim that the bill did not have enough votes.[36]

How could the ESWG expect influential politicians to collaborate, while threatening their power? The ESWG perceived movement and institutional actions as means to build power and pressure on the political elite. From this perspective, contentious movement narratives boosted

Figure 9.2. Ecosocialist working group members protesting with "bribery posters" outside the state government's headquarters in Manhattan, July 2021. The protest happened shortly after powerful leaders in state government had tricked the Public Power NY coalition into failure. The protest is the beginning of the escalation of the "Public Power" campaign that followed. (Public domain photo from NYC-DSA archives)

the power of an organization with little institutional representation but a cause that the larger public cared about. The Public Power NY Coalition conducted a survey indicating that the majority of citizens in the state supported BPRA. Incidentally, the Pew Research Center has found that 67 percent of Americans prioritized the development of alternatives to fossil fuel consumption.[37]

However, the combination of movement and institutional actions carried the risk of collateral damage. This was a concern when group members began discussing the electoralization of "Public Power." The electoralization in 2022 was the most significant expansion of the campaign. It evolved from the frustration with the 2021 legislative session, which left the ESWG contemplating how it could build power to win. The electoralization also developed from a campaign in Queens the year before, when Zohran Mamdani had won a seat in the state assembly on a BPRA platform. The ESWG had participated in protests to close a peaker plant in Astoria owned by the corporation NRG Energy. Mamdani adopted the "Public Power" narrative in his platform: "No Astoria NRG peaker plant. Build public renewables instead!" There was also an organic development in Flatbush. Daniel Goulden, strategy cochair of the ESWG, remembers, "By 2020, I had done a lot of ecosocialist organizing in Flatbush around blackouts and pressuring Kevin Parker." That same year, Daniel and colleagues began recruiting an activist named David Alexis, and they organized Alexis's kitchen cabinet in 2021 in preparation for the campaign in 2022.[38]

The growing number of DSA-backed electeds further motivated the electoralization. In 2020, NYC-DSA went from one to six electeds in state government; in 2022, eight. They worked tirelessly to pass the bill, especially Mamdani and Sarahana Shrestha, sometimes "texting ten o'clock at night to figure out how to get this bill through the budget with all the language and tact that [they] needed."[39] This relationship with a growing number of electeds made the electoralization more meaningful. The success of a movement party culture depends on such close relationships across movement and institutional contexts. Patrick remembers,

> We have a relationship of mutual support where you're not thinking about the elected officials as some separate entity but working with them as comrades. Having elected officials who could help us understand who

to approach and the order in which to approach different politicians was enormously important—who they respond to, what kind of messaging they respond to, and keeping that drum beat up. The electeds did a very good job of supporting the BPRA. There was real dedication on their part. That changed the attitude that a lot of people in the climate movement took towards the ESWG and DSA as a whole. I had people that I've worked with in my professional capacity coming to me and being like, "Well, do you think DSA will support this? Do you think the DSA electeds will support this? Can you put me in touch with so-and-so?" All of a sudden, there's a recognition that DSA is a powerful mover in this space. There are major differences between where we were in 2018 and where we are now in 2024.[40]

While the electeds were supportive in 2021, the ESWG could use more of them, and the group wanted more electeds who put climate first.

The electoralization of "Public Power" was above all possible because ESWG had the internal organization to pull it off. It had become the largest working group in the chapter. Annual surveys asking NYC-DSA members about their priorities show that climate was the priority that grew the most in the period 2017–2021. In 2013, climate change was the top priority for 15 percent of members; in 2017, 14 percent; in 2020, 40 percent; and by 2021, it was the number-one priority. There is a significant increase in members who identify as ecosocialists, as opposed to just democratic socialist or Marxist.[41]

By 2021, the ESWG had become such a force within NYC-DSA that it was capable of essentially taking over the electoral operations for the 2022 cycle. The ESWG influenced the decision to run all campaigns on an ecosocialist platform, and it developed and operated most of the campaigns that year.[42] One of the two victories, that of Sarahana Shrestha, has been particularly important to the group's climate efforts. Patrick describes how the electoralization came about:

We were pushing hard on the legislature in spring 2021 and managed to make the BPRA a litmus test for how serious legislators were around climate. That culminated toward the end of the legislative session in the summer in this major nonviolent civil disobedience outside the legislature's offices in the city. That protest really got a lot of people's attention, but we recognized that this wasn't going to be enough. We had a number of strategy sessions at the Sixth Street Community Center in Manhattan. One was a joint meeting between DSA's Electoral Working Group and the ESWG to figure out how we could electoralize this issue and make it a litmus test for candidates that we were running. It was agreed at this meeting in 2021 that there would be primary challengers oriented around specific pieces of DSA policy, and the BPRA was the centerpiece of that. Nadia Tykulsker and Andrea Guinn played a key role in electoralizing "Public Power" throughout the entire process. I have not seen the climate Left do this before in the United States. In fact, I've never seen people mount a credible threat in that way around climate legislation. With the David Alexis campaign, we were trying to go hard against Kevin Parker on his record, and similar with Sarahana Shrestha and Kevin Cahill. Because Parker and Cahill had such power over the energy system in New York State, they were specifically vulnerable to an attack on their record in that regard. We were also uniquely positioned to point to the electric system in their races, to pin the blame for our existing electric system on the incumbents in a clear and direct way. It was more difficult for some of the other races.

Electoralizing the BPRA made me nervous at the time. If we didn't put up a good showing in Parker's district, then we were finished. But it was the right decision. We wouldn't have won the BPRA if we hadn't been willing to mount those kinds of challenges. We didn't win the Alexis campaign, but we came close enough that Parker could not afford to fuck with us. He couldn't afford to cross us on "Public Power" anymore, or he would risk a bigger and better challenge the next cycle. In the case of Shrestha, the results speak for themselves. Those campaigns were still very alive during the 2023 legislative session. We were pulling out all the stops to pressure different legislators. A lot of my role for 2023 was getting feedback from different organizations and community groups to make sure that we had a broad enough base of support and that the bill reflected the important edits from those sources. I also worked on getting as many cosponsors onto the bill as possible, telling legislators, "This is going to happen, and you have to get on the right side of this."[43]

Michael Paulson, ESWG's cochair of strategy, became campaign manager for Sarahana Shrestha's influential campaign in 2022. His account shows that the politicization of climate hazards in the movement context of "Public Power" was continued in the electoral campaigns. His account also adds nuance to the role of electeds and using the electoral campaigns as a movement tactic:

> PAULSON: Since 2018, all of my organizing in NYC-DSA has been focused on climate. I would say that a major part of the work we've done in 2021 and 2022 is politicizing climate hazards in the city. The climate questions tended to be more about the feasibility of living in certain zones, but it was less about the root causes. With Hurricane Sandy, it was more about the disaster response or lack thereof. People were not necessarily talking about fossil fuels. We politicize these events. There was one day this summer where it was unbelievably hot, and the city texted everyone saying, "We need you to turn off your air conditioner." We had material out immediately with that, and that got huge engagement because it's deeply frightening. It's becoming more common to think of these events as manifestations of the climate crisis, but not in mainstream media—like television media just don't bother.
>
> AUTHOR: How did you get to work on an electoral campaign?
>
> PAULSON: I was not in the market to become a campaign manager or even to work in electoral politics at all. It was entirely an outgrowth of my experience with the absolute disregard for climate change in the state government. We had pushed, we had organized, we had rallied, and we could see there was a certain wall. We couldn't get past a certain point without having our own people in the legislature. And we can see the difference when we had the class of people that were elected in 2020. Having a block, even though it's a small block, they simply are able to challenge the common sense in a direct and aggressive way. The experience of having put so much effort into organizing for climate legislation and to be falling short and just not getting on the radar enough, that made me feel like, "All right, this issue has to become more of an electoral issue."
>
> We knew that it's opaque how laws get made in New York State government, but our tactic was generally canvassing and then trying to identify enough people in the district of the legislature to actively engage and lobby the legislators—in-district, regular people, not

political operatives. That did not work for us, so we reevaluated. We started to add in more in terms of larger demonstrations or protests at people's offices to try to get this attention. Then, this year and last year, we had the most sophisticated strategy, where we simultaneously had an outside strategy based on causing disruption and bad press and general aggravation for people. At the same time, we had a more sophisticated inside strategy of lobbying people, mapping out the different pathways in the legislature, and finding out how to get more access to people who have decision-making power.

Through all of this, we did get to the point where our main bill is the most visible climate bill this year in Albany and in the political press around this issue. But we have not passed it yet. So I would say the culmination of that strategy is we're running a whole slate of candidates on this issue.

We can aggressively say, "You've had this time to respond to the climate crisis with legislation. You have done nothing. You've done absolutely nothing. People are going to die from this. It's very real." That is a pressure that all these politicians have to respond to—the challenge to their livelihood through primary elections. That's where it's come to now. But it's always a diversity of tactics. It will be just adding another layer to the strategy with the electoral engagement.[44]

The electoralization involved the translation of "Public Power" into the organizational context and communication genres of electoral campaigns in local districts. There were people in all six campaigns who identified with "Public Power" and got more involved because of it. "Public Power" also created a new perspective in the struggle against establishment Democrats. However, the translation into electoral campaigns was underdeveloped. The campaigns of David Alexis and Samy Olivares, for instance, hardly communicated the "Public Power" context to its canvassers and constituents. "Public Power" was barely, if ever, mentioned in the canvassing introductions I attended. The campaign management could have said, "The climate crisis is serious. We began exploring solutions back in 2017. We came up with the vision of a government-led decarbonization of the energy sector. We wrote a bill to initiate this process and built a coalition around it, but establishment Democrats in government tricked us it into failure last year, while taking fossil fuel donations! Now we step up the fight. Are you ready?!" Something like that would have helped the many canvassers who did not know this history. The ESWG members did not make this history front and center of the campaign communications (see chapter 6). Also, climate change was not the main grievance of the constituents I canvassed in Flatbush. Most had a more traditional local worldview and did not see climate as the most pressing problem or a root problem. They related to local politicians whom they had known for years, and their main grievances were crime and health care, for instance. Making climate change relatable for more people was an ongoing challenge for the ten to fifteen people who worked on the ESWG's communications team. One of the coordinators, Julianne Feaver, explains,

> The biggest goal of the ESWG's messaging was to make energy and specifically publicly owned energy something that regular people will give a shit about. It's hard to do that. We had to connect why publicly owned renewable energy affects our day-to-day lives and the climate crisis. The climate crisis is so massive and hard to fully process, so we constantly have to bring things back to people's everyday lives. We're spending so much time trying to feed our families and keeping a roof over our heads, you know, fulfilling basic human needs, so thinking about how the climate is changing can be a secondary or tertiary concern for a lot of people. Eventually, our perspective and talking points became some of the main talking points for the people and media that we were targeting, and that's how we influenced how people think.[45]

I observed how some canvassers were able to translate the "Public Power" narrative better into canvassing than I was, mainly because they had more local knowledge and more experience in canvassing. One of them is Michael Pollak, who canvassed extensively in the 1980s and again with Bernie Sanders and NYC-DSA since the mid-2010s. Pollak explained to me how climate worked as a great conversation piece when he canvassed for the Alexis campaign, specifically among the Caribbean diaspora. His account also provides insight into the art of canvassing. Pollak finds that constituents have little interest in politics and little belief that "government can do anything," but he believes that the one-on-one encounter at the door can change this attitude:

With David, oddly, my pitches have mostly been for climate change. I basically say, "Look, so climate change, you care about it?" Generally, it's thought of as a middle-class, white issue. But it turns out, West Indians care about climate change a lot because their islands are in the path of hurricanes, so they think about this all the time!

The whole shtick becomes obvious. You go, "Okay, look, I know how you feel. There's nothing you can do. But we have a plan. First, you take over the power companies, and then we make them do what we want to do, which is convert to renewables and keep your bills down. We make it work in New York, and we make that the model for the US, just like the New Deal. And if you get it in the US, well, then you take it all over the world, right? And you can do that all just by electing your local person. Furthermore, the guy that we're running against is a big guy [Kevin Parker]. He's the biggest obstacle in this way. He's paid lots of money by the fossil fuel companies. Yeah. And he's a powerful committee member. If we could knock him off, everybody would be terrified. So your vote would have a huge effect." Yeah. It's a great shtick.[46]

Why was the "Public Power" narrative not fully incorporated into all of the electoral campaigns in 2022? They were run by ESWG leaders. Did they assume that everyone was now familiar with the narrative? Were they cautiously holding back because climate is still a secondary issue for many people? Whatever the reason, the relative absence of the "Public Power" narrative in some of the campaigns is striking. It is striking from the perspective of the movement rationale for the campaigns and NYC-DSA's grassroots pedagogy of making volunteers "understand the strategy and the political context" to boost their political agency (see chapter 8).

The electoralization also contributed to a transformation of the image of the "Public Power" campaign. The image transformed from an urban movement style of the 2010s to a more mainstream style of the 2020s that still had movement elements:

The previous "Public Power" identity is more red, uses imagery from lightning bolts and things like that. It was developed in early 2021. We decided later that year that we needed a platform that could cover both

BPRA and the electoral campaigns. We came up with the "Green New York" design system. We wanted to communicate our positive ecosocialist vision of the future, incorporating windmills, solar panels, greenery everywhere, and public transit. We also wanted a New York attitude that encapsulated the entire state and wasn't just focused on New York City. One of the main designs has an image of a beautiful Hudson Valley landscape that has trees and hills and rivers with windmills, and then it flows into a New York City skyline with the Statue of Liberty in the forefront. We use the Statue of Liberty to convey a sense of hope for a better future, and it's also a New York icon. We also took visual cues from past movements that weren't limited to climate. We took inspiration from the Black Panther Party and use the font from signs at the March on Washington in 1963.[47]

Just as climate professionals in the ESWG shaped the political project, advertising professionals shaped the communications. Julianne Feaver describes how the communication team's leaders took learnings from working in corporate America and applied them to movement politics. This shows in the conception of the audience. The team primarily targeted "regular people," broadly conceived as people with a casual interest in politics. They did not target "socialists who are constantly thinking about politics and reading the great leftist thinkers of our time, because most people don't have the time, or the mental energy, to do so." The team also adopted methods from corporate public relations for the next steps in the process. It conducted messaging workshops and adopted a storytelling model that "starts with a common ground and then frames the problem and solutions around that." Inspired by this model, says Feaver, "we started our message with universal truths and things that bring people together, like the belief that everyone deserves energy, clean water, and clean air, and connecting the struggles that we're fighting for with people's everyday lives." A key aspect of this messaging approach is hope. The role of the team was to motivate and keep people excited and energized to fight, so they were deliberately avoiding doom narratives. It also followed the principle of omnichannel marketing. "We worked to make our message hit our audiences from all sides and constantly remind them. It's like advertising but for a good cause!" says Feaver.[48]

Timeline of the Public Power (PP) project of NYC-DSA

	2016	2017	2018	2019	2020	2021	2022	2023
Federal	Trump wins			Green New Deal res.			Inflation Reduction Act Pressure from the Squad Bowman in NYT Green Schools rally	
State				Climate Leadership and Community Protection Act				
ESWG		ESWG founded	"PP campaign" > "PP Coalition" with members of the Energy Democracy Alliance				ESWG leaders in electoral campaigns	
Coalition partners								
Legislation				BPRA is developed and shopped around		BPRA rejected	BPRA rejected	BPRA passed
Protest				Power outages and utility bills agitation	Astoria power plant protest		"Bribery" protest activists arrested	Kevin Parker adopts BPRA
Electoral					Mamdani campaign	Decision to make BPRA center of primaries	Six ecosocialist electoral campaigns	
					PP electoralized			
Seats in government			1		6			8

Figure 9.3. Timeline of the "Public Power" campaign. The Ecosocialist Working Group organized internally for years before launching the campaign, and the campaign evolved with the escalation and electoralization enabled by the growing number of electeds and developments in the national political environment.

Conclusion

So, this is my story of how New York socialists won BPRA. It is impossible to distill the lessons into a simple recipe from this history because much of the success came from the capacity to respond to an evolving process that involved a particular movement, coalition partners, and legislative developments at the state and federal level. If I should try, I would say that the success was made possible by the critical mass of talented activists, some of whom had prior movement experience and professional networks, in a movement organization that provided the political independence to propose radical change and, later, the electoral organization and institutional platform to build power, culminating in US House Democrats urging the governor to pass the bill.

Major challenges remain, however. The first is that BPRA is a small step and that such changes might not be at the scale that can save us from climate collapse. Also, the implementation is going slowly, although the government has implemented some aspects. The bigger picture, however, is that it has not substantially changed the course of the state's neoliberal capitalist order. This would require a bigger movement, greater efforts from congresspeople, and the further organizational development of DSA.

This brings me to the second point, namely, that the NYC-DSA's goal of building movement capacity in the climate crisis is difficult without developing the organizational and institutional infrastructure further. The electoralization of "Public Power" helped NYC-DSA secure a political victory, but the organizational capacity that had been developed since 2017 was drastically punctured after the elections in 2022. The ESWG temporarily transformed itself into an electoral organization, channeling most of its resources into developing and running electoral campaigns, and this exhausted the group's resources. It did not help that four of the campaigns lost and that the COVID-19 pandemic had raged for two years. NYC-DSA membership hit a low point at around two thousand in spring 2023, and the ESWG was almost completely exhausted by the time BRPA passed in 2023. One leader told me that the group started pretending that it had more resources than it did and would not have been able to continue the fight if the bill had not passed. Movement-based parties such as NYC-DSA can prepare better for electoralization processes in the future, but they cannot build much more long-term capacity without salaried leaders and more people in government. In the current mode, NYC-DSA is somewhat dependent on recruiting from moral shocks: In the months after October 7, 2023, four thousand new members joined NYC-DSA, and leaders are now working to revitalize the organization and strengthen membership retention.

This is the moment when some critics would claim that NYC-DSA is a failure and that it demonstrates the weakness of movement party organizations in general. I disagree. Scholars and organizers can learn from NYC-DSA's constructive approach to political organization, constantly evaluating successes and failures to make small but significant changes in alignment with fundamental goals. The problem with rejecting NYC-DSA or movement party models altogether is that doing so easily leads to the rejection of the fundamental principle of building bridges between grassroots democracy and political institutions. Such impatient disruptive radicalism is dangerous, especially in the current crisis of democracy. It is a symptom of both the crisis and a lack of knowledge about political organization. Disruptive radicalism is counterproductive to the potential of democracy as a constructive process, and it distorts reality: Small changes in the outcome of NYC-DSA's electoral campaigns in 2022 would have made a difference. Two of the campaigns were narrow losses, and winning those campaigns would have expanded the organization's institutional platform and built capacity for the ESWG. One of the candidates was a cadre ESWG member. This outcome would have made it easier to talk about incremental change and validate the organization's principles. My argument is that the core principles are sound and are relevant to repairing democracy, especially in the context of the local Democratic Party organization, but there is room for improving some aspects to strengthen long-term capacity. There are major barriers to doing so, including dark money and the two-party system, but again, those problems do not prove NYC-DSA and movement-based politics wrong.

Conclusion 1

Status of the Democratic Road to Climate Survival

How much has NYC-DSA advanced the democratic road to climate survival? The cynical skeptic might burst out, "NYC-DSA has changed little. The political situation in New York State has not fundamentally changed, and the speed of the state's energy-sector transformation falls short of being meaningful. Also, the wider national and international democratic socialist movement has partially collapsed."

I am going to challenge this interpretation, although I do not deny that it seems compelling at first glance. The problem is not that its individual elements are wrong, but they are lacking a perspective for how political change is enacted democratically. If we are simply seeking to judge emerging local movements such as NYC-DSA, we are not stimulating thinking about how these attempts to rebuild democracy can be advanced further. At worst, a simplistic evaluation of the electoral and institutional achievements can be an argument for throwing in the towel. That would bring the climate Left back to square one and lend power to neoliberal fascism.

A central aim of this conclusion is therefore a balanced evaluation with a perspective for the next step in the democratic road to climate survival. I take inspiration from socialist organizing and movement leadership by focusing on how the movement can build on its achievements, emphasizing the idea of democracy as a constructive process across movement and institutional contexts.

Ten years on, what did the US democratic socialist movement achieve besides a few electoral victories? The movement articulated new political interests that continue to structure political struggles. The Occupy Wall Street movement changed the way we think about capitalism, and the climate movement turned climate into a matter of public concern. The democratic socialist movement developed these interests into electoral campaigns, including the first socialist in presidential primaries since the 1920s. In New York City, it won climate legislation. The New York State Build Public Renewables Act is the product of collaboration between socialist movement leaders, electeds, professionals in the climate field, and progressive Democrats. The socialist movement played a decisive role in this process. It changed how climate politics is written in New York.

This leads us to another main achievement, namely, the development of a new form of political organization in neoliberal capitalism. NYC-DSA evolved into a movement-based organization with party elements and capacity for long-term organizing. It created the movement party environment in which socialists and climate professionals could develop BPRA. BPRA could not have been developed in the Democratic Party. The six-year process that culminated in the passing of the BPRA shows the organization's long-term capacity and movement party dynamism. The synergies between movement and institutional contexts, including the translation of BPRA across these spheres were crucial to winning BPRA. The "Public Power" campaign began in the movement, created a large coalition across the state, pivoted to a new level through electoralization, and finally won with the support from socialists in Congress. However, NYC-DSA is struggling to reach its goal of building long-term organizational capacity in the climate crisis, because of the exhaustion after the electoralization of BPRA and the pandemic.

NYC-DSA's approach to political organizing holds significance to wider national and international efforts to rebuild democracy. The key potential is the combination of an explicit political project with a movement leadership approach to community organizing. Socialist demands for social justice and democratization of the political process have become more urgent with the rise of Trumpism. Trumpism exploits and amplifies the crises of climate, inequality, and democracy, bringing society toward collapse. We have seen that NYC-DSA is not without divisive affective and moral commitments such as contempt and disgust, but it is not consumed by these feelings. NYC-DSA's leadership conceives democracy as a constructive, grassroots-based process. Participants are not confined to promoting a campaign but are developed into reflexive political agents in democratic processes. They shape the political project.

NYC-DSA further uses history as a mobilizing resource. It does so by pointing to the achievements of movement-led socialist coalitions in the past: the welfare and labor reforms of the 1930s and the civil rights legislation of the 1960s. The underlying argument is that socialists can win again, and this time on climate too. It is a monumental challenge, but if we adopt NYC-DSA's approach to building long-term capacity, the question for the democratic socialist movement is not, "Did we win or lose in one decade?" Instead, the question is, "How can we build on the victories, organizational capacity, and experiences of the 2010s in the coming years?"

This book has shown that NYC-DSA is a movement shaped by the twenty-first-century US metropolis. It adapts democratic socialism to the US political system, and its leaders have diverse backgrounds and eclectic inspirations within and beyond the city. They come from presidential campaigns, tenant organizing, radical far-Left politics, queer politics, and the climate movement. NYC-DSA integrates democratic socialism with racial and gender justice. Future research can usefully explore its queerness, which I think has helped push back against the toxic masculinities of past socialist movements. Indeed, other political organizations can look to NYC-DSA for inspiration in the organization's cultural dimension, not just its structural design. This is also relevant because political organizations, including movement parties, are contingent on cultural values and practices, not just generic models.

Is the Situation Getting Worse? What Can Ecological Socialists Do Next?

Let us turn to the national situation and recognize that the situation is getting worse, that climate change is getting worse and democracy is dying. Does this diminish the value and potential of the democratic socialist movement?

Popular conversations about the political situation tend to focus on which party has the presidency. Surely, the choice between Kamala Harris and Donald Trump in 2024 presented different paths for the future. But having a Democrat president will not immediately bring an end to the crisis of democracy. Democrat presidents over the past two decades have not created major changes to the political system. Trump undermines democracy, but he too evolves from a longer history, from the long ultra-Right movement. As of spring 2025, there are signs of a Sanders–AOC resurgence with a more universalist appeal beyond Democratic Party affiliation. This is also an attempt at surviving the widespread collapse.[1]

By 2024, four major interlocking crises had spiraled for years:

1. *The crisis of inequality.* Wealth inequality has been rising for sixty years and is now higher than in almost any other developed country. Forty-four percent of Americans are not making a living wage. Record-high rents are causing record-high homelessness.[2]
2. *The climate crisis.* Global carbon emissions remain at an all-time high, and global temperatures exceeded 1.5°C above preindustrial levels in 2023. The speed of temperature rise requires new prognostic models and has thrust the scientific community into a state of despair. Climate hazards such as flooding and wildfires are happening more frequently, quadrupling the annual cost of such events over the past forty years to $150 billion in the United States alone. The IRA of 2022 was a historic legislative climate effort, but it also included permission for the expansion of fossil fuel production, and the second Trump administration is now rolling back climate commitments.[3]
3. *The crisis of democracy.* The democratic problem of the economic elite's influence on the Democratic Party is bigger than ever. It was a major contributing factor to Sanders's loss in 2016 and the biggest blows to democratic socialism since then, namely, the loss of Jamaal Bowman and Cori Bush in the 2024 primaries. Bowman and Bush were the two most outspoken critics of Israel's genocidal war on Palestinians in Gaza and became the target of the Democratic Party's pro-Israel lobby. The American Israel Public Affairs Committee (AIPAC) spent a record amount of money on ads targeting these two candidates—their opponents had the biggest budgets ever in primaries. The Democratic Party helped the pro-Israel lobby. It gerrymandered Bowman's district to incorporate a suburban area with a high concentration of pro-Israel Jews and recruited

George Latimer, a seventy-year-old pro-Israel politician from that suburb, to run against Bowman.
4. *Global militarization.* Putin's war on Ukraine has led to a global militarization that sidelines climate. NYC-DSA has attracted many new members through its protests against Israel's genocide in Gaza, but this war has also created a more challenging political environment for socialists, as demonstrated by the fate of Bowman and Bush.[4]

Classic movement theory of mobilization stipulates that people will participate in collective action if they are experiencing unacceptable injustices and find the protest narrative meaningful. But what if there are too many crises and too many narratives? We live in an age of both communication overflow and social disintegration, including a fragmented media landscape. Media communications are simultaneously informing, entertaining, and scaring us. It requires a deliberate effort to avoid becoming digital couch potatoes. But political journalism is written for consumers with a busy workday and little time to fight the unprecedented disaster of climate collapse. Through fieldwork, I learned that New York socialist organizers look at the world differently. They focus on a political project and use a hybrid organizational approach to win small victories with a larger goal. They identify predecessors and learn from them. When they look at major crises in the past, they do not focus on the negatives. They look for the methods that led to progress and the evidence that progress is possible. They know that major social justice reforms have been possible in recent history even when the challenges seemed insurmountable.

The ESWG's victory with BPRA shows that a critical mass of talented people can influence climate politics and bring us meaningfully closer to climate survival if they are highly committed, learn from experience in diverse organizations, develop a political project, build coalitions with climate professionals, and exploit opportunities in the political environment.

The current political situation in the country is extremely challenging for the climate Left, but the Left can learn from NYC-DSA's focus on strengthening existing structures and growing collaborations between grassroots movements and electeds. This approach remains essential to a just ecological transformation, especially in the crisis of democracy.

For example, the electoralizing of "Public Power" exhausted the ESWG, but it would have been easier for ESWG to regroup from the election if a few more campaigns had led to victory. More victories would have created wind at the back, as well as salaried jobs for some of the organizers and candidates in government.

Adding to the list of achievements, NYC-DSA has helped thousands of people overcome the negative spiral of political participation, reintroduced grassroots democracy in the city, and overcome much of the factionalism of past leftist movements, while also extending their reach through electoral politics. It also helped revitalize political citizenship for young people in an era of growing antidemocratic tendencies.

The main weaknesses are the difficulty in sustaining the movement in the long term without a larger institutional base, plus the high turnover of organizers. The organization could gain strength if it became more institutionalized in economic life and could make a difference in the material conditions of members. In its current form, NYC-DSA to some extent relies on young idealists with time for activism, and it recruits most members during moral shocks, such as Trump's election and the genocide in Gaza, where the Democratic Party's moral failures are pronounced. The organization's party dimension has been more stable and experienced more linear growth.

Can There Be Climate Justice Without Socialism?

The democratic socialist movement has challenged the turn away from ideology in neoliberalism, the idea that ideology is dead. The movement has used the word "socialism" but has deliberately focused on concrete issues and not the label. The same can be said of the new NYC-DSA, which focuses on explicitly political collective action. This return of ideology in search of political justice in late neoliberal capitalist society has also found an expression in academic writing. The most visible trend is the revival of Marxism, but there is also a subtler change in thinking. Many scholars are deeply moved by the crisis of inequality and climate change, and this shows in their thinking. My encounter with NYC-DSA opened up new ways of thinking about socially relevant scholarship. It led to a transformation of my sense of purpose. My aim is not to produce scholarship for professional reasons alone, because I do not think I can afford to do so any longer. I have to try to advance democratic climate solutions. The climate crisis puts the institutions of knowledge production in a new light and hopefully changes the criteria of relevance. I offer some suggestions to this end in the second concluding chapter.

This book expands on the formative argument in climate sociology that climate change is political. I have argued that decades of inadequate government action on climate have led to a feeling of neglect that complicates the climate movement's relationship with politics. The movement is increasingly politicizing climate change, however, and it is increasingly influenced by emerging Marxist analyses of the fossil fuel industry's power, the biased media coverage of protests, and the fixation with economic growth in institutional politics.

Ideology is crucial to the basic questions in political struggles over climate change. What does it mean that the elites have disproportionate influence on the political process? How should the government protect communities hit by floods and wildfires? Who will live, and who will die? Should polluters pay? The neoliberal solution is to expand fossil fuel production and let corporations profit from climate hazards. The democratic socialist solution is a faster transition to renewables, more democratic ownership and public infrastructure, and greater justice for vulnerable communities.

A discussion of socialism thus helps clarify alternatives to the dominant neoliberal order. The negation of ideology helps the ruling class escape accountability. It is a gaslighting strategy that serves the interest of reckless capitalists and climate deniers.

Socialism is built into the structure of the political field and the history of environmentalism. The struggle for the environment has always been fought by the environmental movement and the Left. The environmental movement began with the scientist Rachel Carson challenging chemical corporations in court and evolved into leftist green parties and ecological socialism. The Republican Party is still deeply embedded in the anticlimate movement. Internationally, the situation has become more blurred. Some right-wing movements now appeal to workers and recognize climate change. This happened in France, for instance, with the retooling of the National Front in 2018. The party adopted the more welcoming name Rassemblement National (National Rally), meaning "national gathering," and a nationalist and nativist climate politics. It has attracted lower-income rural voters who traditionally supported leftist parties. However, National Rally has recently abandoned its welfarist promises and developed closer ties with the economic elites, and it does not give high priority to climate. The latter can also be said of neoliberal social democratic parties. That leaves democratic socialists in the position of providing an alternative to neoliberalism and fascism. Democratic socialism has been powerful in the past and could become powerful again. To this end, democratic socialists must claim a right to self-definition against Cold War stereotypes and build their vision for a just ecological transformation of society.

Conclusion 2

Scholarship for Survival

I have been transparent about the transformative experience that shaped this book: how I discarded a research project in favor of working on the climate crisis and getting involved with a socialist movement. This experience has changed my perspective on scholarship. I made a methodological argument for ethnography in chapter 1, but I would like to push the envelope further. I think social scientists can do more to advance the democratic road to climate survival, and I am suggesting not only more scholarly activism but more engagement in exploring political solutions. Doing so requires elements that are central to this book, including ethnography and movement leadership, but it can be advanced at the collective level by reorienting social science disciplines. This will necessarily include a conversation about what the primary goal of social science is. To help save the planet we should not only produce more climate research. We should move beyond individualistic, career-driven scholarship and neoliberal politics. We should embrace socially engaged research and help shape next-generation political leadership and organizing around the explicit political goal of climate survival for all. I am calling for a movement in sociology toward scholarship that is acutely aware of the climate crisis, engages closely with citizens, and contributes knowledge to guide the democratic road to a socially just climate future.

I start by challenging the institutional logic that keeps scholarship on political organization stuck in an old worldview and detached from climate. There is an international wave recently of movement scholars calling for bridging the gap between movement and party research. This scholarship shows how movements are transforming political parties now. However, the arguments are still focused on organizational forms as structural solutions within the political field. One aspect of the dominant logic is the idea of political organizations as solutions to practical problems of mobilization and social choice. This idea is illustrated by Kitschelt's argument that movement parties emerge as solutions to the questions, "How can citizens be mobilized, and what can they agree on?" This somewhat functionalist and technocratic approach to organizational survival is also relevant to thinking about climate politics, but it fails to address the more fundamental challenges that political institutions are facing in the intertwined crises of democracy and climate. In a word, scholarship of political organization could think more about climate survival. How are political organizations mediating the interests of citizens and institutional actors in the climate crisis? How are political parties embedded in fossil fuel capitalism, and what are the alternatives? What are the analytical parameters and emerging forces in the planetary crisis that we need to reckon with to avoid the fate of Steinbeck's tenant farmers—being trapped by outdated solutions and going extinct for that reason? How might the insights of ecological economics and Earth systems science, for instance, inspire new theory of political organizing?

Analytically, ethnography is essential to this challenge. Ethnography investigates the smallest unit of political action—the individual citizen—and at the same time incorporates the broader perspective of how the political field relates to society and the climate. It can integrate analytical perspectives on individual agency, national political institutions, economic inequality, and climate change. If we consider only one of these perspectives, we will have a more limited understanding of political culture and how people are experiencing it. In the following, I illustrate this point by addressing the evolving problems of two fundamental dimensions of modern democracy since its formative years after the French Revolution: media and capitalism.

The Media Environment of Politics

Media have played a central role in democracy from the beginning. Newspapers and magazines created national spaces for sharing information about societal matters and for interpretation and debate. The national circulation contributed to the concentration of power in national unions and parties. In the twentieth century, radio and television expanded the sensory dimensions of media into sound and moving images, and live broadcasting defined political events for audiences as they were happening. Media rituals of celebrating national leaders and political events emerged. Mass media also boosted fascism and the culture of political celebrity. They eventually contributed to the decline in embodied political participation. Citizens would now watch political events on television and feel less of a need to participate in local party meetings or in collective action. Citizens were disembedded from local political organizations and reembedded into national media.

Those media dynamics are evolving and transforming in the hybrid media environment of mass media and digital technologies in the early twenty-first century. Digital media are fast and dynamic, constantly in motion. They transform social relations and geographies through an expanded range of connectivity and through large social media platforms that are challenging the sovereignty of traditional media institutions. Digital media, moreover, allow more people to produce and share information, and this has democratizing effects but also contributes to misinformation and information overload. In personal life, they are accelerating a broader trend of individualization through personal devices and algorithmic recommender systems. The development of networked personal media has created the social media influencer, which represents the expansion and destabilization of the institutional framework of media. In the political field, institutional media still have much power. The ever-increasing advertising budgets for presidential campaigns show that most of the money still goes into TV ads.

However, the deeper and accumulated impacts of media developments are changing the conditions of political life. Social media platforms help ultra-Right movements spread misinformation, win elections, and reward bullying autocrats. The overload of mostly negative information makes individuals feel smaller and less powerful, thus impacting the psychology of political agency. When media increasingly decide what is deemed important, they are compromising a more holistic understanding of the world through embodied social relations. The world "in the media" seems bigger and more powerful than the locally experienced world does. We rely more on the media communications of political parties and celebrities, even as we complain about them. We find it exotic when canvassers knock on our door and do not replicate a generic message from a mass-media campaign. Meanwhile, the social basis of local political institutions is eroded by translocal flows of power and the increasingly translocal problems.

My experience in NYC-DSA made me aware of the psychological impact of media and gave me a new sense of agency. Local political participation can help balance negative implications of media and build relations of trust, but not if the organization just uses participants as vehicles from a narrow media logic of promoting a celebrity candidate. It makes a big difference if participants can have influence and work together as equals. My field research also taught me that local grassroots organizations can draw from translocal flows of climate thinking to influence climate legislation in a way that parties relying on mass media and corporate donors simply cannot.

The Organizational Environment of Politics

Like national media, the main institutions of the nation-state have existed for two hundred years. In this section, I argue that political parties are being transformed by their larger organizational environment of neoliberal capitalism.

In general terms, organizational forms mediate collective interests and articulate value systems. In Indigenous societies, clans and tribes mediated the interests of small local groups that were highly dependent on their immediate natural environment because of limited means of transportation and communication. In feudal society, the church and aristocracy served an elite of landowning lords, their power partly secured by limited education and technologies of communication for ordinary people. Modern democratic societies evolved from political parties mediating collective interests, and these institutions exist within the system of capitalism. The expansion of politics beyond the institutional world of parties that began in the 1960s was the symptom of a wider transformation of traditions and class structures, with new emancipatory cultural demands. Capitalism has since become deregulated and transformed politics and media. Today, we live in a new Gilded Age of extreme inequality dominated by global financial markets and corporations. The deregulation of financial markets and news media and the weakening of antitrust laws have led to the creation of larger corporations. Tech firms in California, Wall Street banks and asset management companies, and media firms such as Fox and News Corp are more than just businesses. They create a normative organizational environment defined by managerial, legal, and technological infrastructures and by monopoly. Public-sector and civil society organizations are adapting to this environment, thus transforming the rights and role of citizens and the outlook of democracy. What does this mean for political life? Party organizations become more like corporations, more centralized and technocratic, and their communications look more liked that of PR agencies—it is all rhetoric, disconnected from material realities. From this perspective, one can understand why movement culture appeals to the need for a simpler, more relatable, and personalized form of politics and, in the case of socialist movement parties, a more just and substantive political culture. Finally, the normative and material power of monopoly makes it more difficult to mobilize support for alternative futures and independent organizations.

A specific implication of this new Gilded Age for the political field is the rise of oligarchic tendencies and dark money. This too has been boosted by deregulation, particularly with the deregulation of campaign finance law after the *Citizens United* Supreme Court decision in 2010. The news is not that conservative interests have more money than leftists. The news is that the corporations and wealthy individuals can organize more in the dark and have moved further right.

This picture emerges from Jane Mayer's book *Dark Money*. The book shows how at the beginning of the twenty-first century, the Koch brothers created a network of billionaires to promote free-market ideology and then to oppose Obama. More billionaires participated in this network than existed in 1982. It included the likes of Supreme Court Justices Antonin Scalia and Clarence Thomas. The amount of dark money in politics more than doubled after the *Citizens United* decision. It exceeded $1 billion in the 2020 election cycle, and much of it went into the Biden campaign, which had a total budget of $1.6 billion. Dark-money groups also donate hundreds of millions of dollars to politicians in Washington outside elections, resulting in growing corruption and declining trust. Dark money thus accelerates the vicious cycle of participation.

The political scientist Jeffrey Winters argued in 2011 that democracy in the US had become a form of civil oligarchy, a political system dominated by the wealthy few. Trump 2.0 has taken oligarchy to the extreme, demonstrated by the unprecedented role of the world's richest person, Elon Musk, as the primary funder of Trump's campaign and then securing billions in federal contracts and leading the overt slashing of government agencies. Winters points out that oligarchs have always been part of US democracy and that oligarchs consistently redesign democracy when it overperforms. The solution, he says, is organizing from below during crises and having long-term organizational capacity. The New Deal reforms were possible because unions had mobilized for decades before the Great Depression. When the Great Recession happened in 2008, unions were much weaker. Critical infrastructure was lacking. The bailout was on the terms of the oligarchs. From this perspective, we should muster patience and endurance to fight for justice and climate survival. Giving up on building long-term capacity would result in a sure loss. We would stand to lose everything.[1]

Scholarship for Survival

I agree with Winters that mobilization from below is needed to create a more democratic and socially just society. But how exactly can this be done, and why is movement scholarship not addressing this question more? The academic literature on movements is primarily motivated by an interest in explaining rather than in guiding political change. It could be more proactive and think more *with* rather than *about* organizations that work for democracy and climate.

In this urgent crisis political research can still serve the function of creating clarity of the big picture. Such knowledge can help society steer through complex and emotionally difficult situations. But political research now has the responsibility to move further beyond building disciplinary knowledge and coproduce knowledge with political entrepreneurs for democracy and climate. Scholars can learn from the way movements think to create hope in research and education and adopt constructive approaches to democracy to advance this movement project. Research can help identify successful approaches to political organizing and clarify expectations and goals. It can help people expand the perspective from intuitive thinking and panic to the long-term organizational processes necessary to create structural change.

I do not call for a revolution but a change in how scholars understand their role and some disciplinary reorientation. I have addressed the role of scholars earlier and will only add that we should learn from the problems with career politicians and not idealize career scholars. The disciplinary reorientation could involve the introduction of disciplinary perspectives of leadership and strategy into political sociology. I propose a combination of three perspectives: The first is movement leadership, the potential of which has been demonstrated by Marshall Ganz and Jane McAlevey. Their ideas have proven useful to many civil society organizations over the past few decades. Ganz and McAlevey have looked beyond the conventional discourse of movement scholarship, drawing from their experience in organizing and incorporating elements from other disciplines. Their thinking is not primarily guided by an interest in refining a theory or discipline but by understanding how intentional social change can be achieved. Ganz has taken inspiration from positive psychology to think about how hope can be mobilized and channeled into collective action. He builds a bridge to broader conversations about leadership and management. McAlevey adopts a more critical approach to how unions can protest and negotiate strategically in labor struggles with corporations.

I have argued in this book that the microlevel movement leadership perspective needs to be complemented by thinking about organizational ownership. This brings me to the next disciplinary perspective that scholarship on political organization can usefully incorporate, namely, the political economy of organization. My inspiration for this perspective comes from the realization of the way Ganz's approach has been appropriated into neoliberal agendas. It lends itself to this appropriation because it emphasizes positive psychology and does not explicitly address politics or critical thinking about power. My thinking about the political economy of the political field has obviously also been inspired by NYC-DSA, with its emphasis on independent ownership, grassroots democracy, and struggle against the Democratic machine. A high level of political and economic independence is vital to local democratic culture.

Finally, organizational ethnography is necessary to produce holistic understandings of how organizations work to advance the democratic road to climate survival. We need to know more about this to help inspire and demystify this form of work. We can show how citizens become political organizers and entrepreneurs and what they do to effect change and build long-term capacity. Ethnography can show how movement leaders collaborate with activists and climate professionals in organizational processes. It can capture the informal aspects of movement culture and help balance media representations of political culture.

Together, the disciplinary perspectives of movement leadership, political economy, and ethnography can bring scholarship on political organization further with the goal of advancing the democratic road to a socially just climate future. The individual perspectives have been developed, but the combined potential has not yet been exploited. Let's go!

ACKNOWLEDGMENTS

This book would not have been possible without the collaboration of many people in New York political life. I thank dearly all of the informants who appear on the pages of this book. For their support at vital stages in the process, I owe special gratitude to Abraham Silverstein, Cihan Tekay Liu, David Duhalde, David Alexis, Emma Buretta, Gustavo Gordillo, Hunter Rabinowitz, Isabel Anreus, Jack Gross, Jerry Krase, Julia Salazar, Michael Paulson, Patrick Roberts, Shay O'Reilly, and Tascha Van Auken.

Sarah Reibstein read the first complete draft in early 2024 and worked with me on important issues. A few months later, I received two reviews from NYU Press. The invaluable suggestions from Sarah and the reviewers led to major improvements. Thanks also to Andrew Katz and Paulina Cossette for impeccable copyediting at various stages in the process. Any remaining weaknesses are my responsibility, of course.

Ilene Kalish at NYU Press is one of those rare academic editors with scholarly attention to substance and a remarkable talent for translating projects into the world. She made invaluable suggestions and helped make the final stages of the process the best possible experience.

I am very grateful to Eric Klinenberg for inspiration and being a wonderful host at the Institute for Public Knowledge at New York University.

The book benefits from conversations with great colleagues at Roskilde University, including Elisabetta Petrucci, Eva Mayerhöffer, Louise Phillips, and Pernille Almlund.

This book was partially funded by a generous one-year fellowship from the Carlsberg Foundation (grant number CF20-0292). The fellowship allowed me to experiment, take risks, and explore a new disciplinary path.

Finally, I am especially grateful to my wife, Anja, and our children, Louisa, Dante, and Victor. You bring light in the dark times.

NOTES

INTRODUCTION

1 It is difficult to measure the casualties of global warming because the consequences of changes in global ecosystems are complex. It is also difficult because vulnerable populations have less power and are therefore less accounted for. The World Health Organization (WHO) estimates that there have been 489,000 annual heat-related deaths since 2000 and that the number of heat-related deaths in Europe has grown 30 percent since then. Climate change is a major driver of population displacement in Africa. For a comprehensive scientific study, see IHME 2021. In 2021, around two and a half million people were displaced due to disasters (World Meteorological Organization 2022a, 6, 29). For a report on the situation in Latin America and the Caribbean, see World Meteorological Organization 2022b. See also Goldberg et al. 2021; Leiserowitz et al. 2022; Copernicus 2024; Carrington 2024; Klinenberg et al. 2020; Kluge 2024.
2 Aronczyk and Espinoza 2022; Fletcher et al. 2024; Coombs and Tachkova 2024.
3 Jackson 2013; A. Brown 2013.
4 Bourdieu 1990.
5 Gessen 2020.
6 Harvey 2005; Laybourn and Dyke 2024.
7 N. Klein 2022.
8 Keith et al. 2023, 8–9. The situation is reflected in financial markets, where the value of climate investment products, the so-called ESG index funds, has declined, while the markets for artificial intelligence, weapons, and fossil fuels are booming. Corporations are backing off from climate targets, and they are trying to bypass climate legislation by making the accounting so complex that it becomes nontransparent and incomprehensible to ordinary people, which is the same approach that they have taken to tax law for a century (Buller 2022; Temple-West 2024; Bryan and Mooney 2024; Bryan et al. 2024; Masters and Temple-West 2024; Pistor 2019).
9 Piccio 2019, chap. 1; Schlozman and Rosenfeld 2019. I distinguish organic grassroots movement organizations such as the Democratic Socialists of America from professionally managed advocacy groups without chapters and members (Skocpol 2003).
10 Kitschelt 2006; Sartori 1976, 63; Goodwin and Jasper 2004, 137.

11 Tarrow 2021, 162–171; O'Carroll et al. 2024; Stetler 2024; García et al. 2024.
12 Talmadge 2023; Elstub and Khoban 2023.
13 Economist Intelligence Unit 2023.
14 Freeman 2000.
15 The understanding of democracy as a protocol for defining common rules in a pluralist society has been pioneered philosophically by John Rawls. In his seminal theory, Rawls ([1993] 2005) argued that the sixteenth-century Reformation was a formative development because the coexistence of two religious doctrines created a social order that accepted pluralism. Rawls praised liberal democracy for creating protocols for a fair negotiation of pluralism.
16 Benjamin 2012, 6–7; Galie 2012, 19–20; World Economic Forum 2018.
17 Hobsbawm 2011, 5, 385. The finance journalists Andrew Ross Sorkin and John Cassidy published book-length accounts of the culture on Wall Street in the years and months leading up to the bankruptcy of Lehman Brothers in September 2008 (Barrett 2009).
18 Kose et al. 2020, 10.
19 Our Revolution was not a democratic membership organization. It was created in the aftermath of the election by some of Sanders's former core staff to further grow the political project and transform the Democratic Party (Duhalde 2020, 2021; Moody 2022, 58–59).
20 Sam Lewis, interview with author, May 1, 2022; Tarleton 2015.
21 Gordillo 2024; Central Brooklyn Branch 2024.
22 DSA's membership numbers include expired members whose last dues payment was more than one year ago but less than two years ago. The expired members generally make up around 15 percent. National membership numbers: 6,200 in 2015, 11,000 in 2016, 32,000 in 2017, 55,000 in 2018, 56,000 in 2019, 87,000 in 2020, 95,000 in 2021, 83,000 in 2022, and 78,000 in 2023 (Hernandez and Huang 2023, 19). Membership numbers for NYC-DSA are not public (NYC-DSA Steering Committee 2023).
23 Day and Uetricht 2020, 59, 80, 92.
24 Mezirow 1978, 104–105.
25 Temple-West 2023; Taylor 2020.
26 I acknowledge the atrocities of both totalitarian communism and imperial capitalism (Sebestyen 2018; Khlevniuk 2017; Snyder 2010; Malia 1994; Kotsonis 1999; Bevins 2020).
27 Stout 2010; McAlevey 2020; Moody 2022.
28 Moody 2022, 1.
29 Chakrabarty's (2009) idea of a new negative era in history helps understand this dark cloud on us.
30 Maslow 1943, 373. On the concept of transformation in the climate crisis, see Holt 2022.
31 NYC-DSA, n.d.-a.

32 S. Lewis 2022. The far-Left offshoots tend to be more proactive in social media polemics than in practical organizing. They include Reform & Revolution and Marxist Unity Slate. Further to the left is the Libertarian Socialist Caucus. There is no official list of caucuses on DSA's website. For an exploration, see Dreyer and Locker 2021.

CHAPTER 1. THE CLIMATE MOVEMENT NEEDS POLITICS!

1 Habermas 1996, 359, 381.
2 *Deutsche Welle* 2024.
3 Adam and Randerson 2010; N. Klein 2014, 12. Fisher (2024) aptly describes the history of the COP conferences as a history of disappointments. Dyrhauge 2020, 20. In 2024, XR stated, "Our government has failed to make the urgent changes required for a livable world, despite the urgency and scale of the climate and ecological crisis" (Extinction Rebellion UK 2024). Hattenstone 2021.
4 J. Klein 2015; Lin 2015; Setzer et al. 2022; Climate Litigation Network, n.d.; Walker 2022.
5 Hildyard and Wolfe 2002, 680.
6 Piven and Cloward 1977.
7 The number of Americans alarmed by climate change doubled between 2016 and 2021: 13 percent thinks it is too late "to do anything about global warming" (Leiserowitz et al. 2023, 4).
8 Fisher 2024; Extinction Rebellion UK 2023, 2024; Amelang 2024; Meaker 2024.
9 Fisher 2024; Malm and the Zetkin Collective 2021; Malm 2021. The critique of movement-centrism in political sociology is addressed in chapter 3.
10 Rawls ([1971] 1999) argued that protesters can persuade but should not actively interfere with the conduct of others. This is civil blackmail, not civil disobedience. He maintains that coercion is illiberal because it constitutes "a final expression of one's case." In his view, protest should operate strictly within the bounds of symbolic action. The coercive turn critiques liberalism for being a form of moralism. Robin Celikates argues that the liberal view fails to recognize power relations, calling attention to the conditions or reality facing resistance mobilized from below against dominance and oppression from above. Interest in challenging "liberalism's moral dichotomy of persuasion and coercion" has led Alexander Livingston to argue for shifting the conversation from debating the permissibility of violence to the uses of coercion. Daniel Markovits argues that coercive shocks can "enhance democracy by bringing urgent issues to public attention and kick-starting democratic deliberation" and serve as a counterweight to institutional inertia (Livingston 2021).
11 Brulle and Dunlap 2015, 3–4, 12; Falzon et al. 2021; Huber 2022a; Aronczyk and Espinoza 2022; Lamb et al. 2020. Climate sociology is both a corrective to the parent discipline and a subfield of its own. It critiques the parent discipline's sociocentrism. See also Banerjee et al. 2015; Mulvey et al. 2015, 1.

12 *Guardian* 2022; Associated Press 2023. I edited the quotation a bit for readability. Another Thunberg statement: "Climate breakdown and inequality are linked together and fuel each other. If we are to overcome one, we must overcome both" (Oxfam 2023, vi).
13 Extinction Rebellion UK 2023.
14 McCarthy 2019; Berry and Sobieraj 2014; Hersh 2020, 16; E. Klein 2020, chap. 1.
15 Jasper 2018, 144.
16 Hersh 2020; Hacker and Pierson 2015; Tarrow 2021, 181–182; Mason 2018.
17 Stockemer and Sundström 2023.
18 Helen Mancini, interview with author, April 9, 2022.
19 Amanda Litman, interview with author, March 30, 2022. Litman continued, "Joe Biden is president now, but when he was twenty-nine, he was a city councilor in Delaware. We need more twenty-nine- or thirty-five-year-old young people of color who are running and winning locally, so that they have time to build careers that can help them rise. It's a pipeline. And you have to fill the pipeline more broadly. I also think that young progressives generally approach politics very differently because of their relation to the Republican Party. If your political awakening or engagement started in the early 2000s, you've never known the Republican Party to be good-faith negotiators. I think that some of the younger members of Congress—AOC, Jamaal Bowman, Sara Jacobs—know you can't count on Republicans to do what they say they're gonna do."
20 De Schweinitz 2015. When other people gained voting rights, beginning with people of color in 1870 and women in 1920, the amendments to the Constitution were not fully implemented, and the cultural repression continues to this day. People of color were eventually granted voting rights, and the voting age was lowered to eighteen in 1971. Senator Elizabeth Warren and Representative Nikema Williams claimed in 2022 that the promise of the Twenty-Sixth Amendment remains unfulfilled.
21 Stockemer and Sundström 2023.
22 Bourdieu (2002) 2008.
23 Alford and Friedland 1985, 250.
24 Writes Chomsky, "Democracy, in the United States rhetoric, refers to a system of governance in which elite elements based in the business community control the state by virtue of their dominance of the private society, while the population observes quietly. So understood, democracy is a system of elite decision and public ratification" (1987, 3). A conservative nonprofit group called Citizens United challenged campaign finance rules and won its case by a 5–4 majority in the Supreme Court in 2010. The so-called *Citizens United* decision means that corporations can spend unlimited funds on campaign advertising if they are not formally coordinating with a candidate or political party (Lau 2019).
25 There is a history of critical thinking about news media from the Frankfurt School and French Marxists such as Louis Althusser to the Anglophone political economy of mass media and cultural studies. Chomsky 2002, 63; Habermas 1989a, 1989b; Fraser 1990, 77; Garnham 1979.

CHAPTER 2. THE SOCIOLOGY OF MOVEMENT PARTIES

1. Coser 1966, 1. There is a wider definition of political sociology that looks at power relations in social life in general and without a focus on the relationship to political institutions. That definition developed in the study of movements that sought to influence the public sphere more than the state.
2. Weber (1921) 2009, 78; Bourdieu 1990.
3. Jasper 2020, 627; Nash (2000) 2010, 87; Janoski et al. 2020, 4–5.
4. Bottomore 1979; Lamont 1994; Nash (2000) 2010; Savage 2012; Bourdieu 1996; Skeggs 1997.
5. W. Brown 2015, 28.
6. Graeber and Wengrow 2021, 7–8.
7. Piketty 2021, 1–2.
8. Skocpol and Tervo 2020; Fisher 2019.
9. Della Porta, Fernández, et al. 2017, 34; della Porta, O'Connor, et al. 2017, 1.
10. Della Porta, Fernández, et al. 2017, 36, 53. The far Right won a quarter of the vote in the 2024 election, so Europeans can easily relate to the situation in the United States, even though many Europeans like to feel smart and laugh at Trump, thinking that a similar situation will not evolve in Europe. The far Right in the European Parliament is divided and therefore struggles to translate electoral seats into power (Mudde 2024).
11. Beck and Beck-Gernsheim 2002.
12. Kitschelt 2012, 150–156.
13. Della Porta, Fernández, et al. 2017, 32. The decline of trust and participation is discussed extensively in Putnam 2000.
14. Giugni and Lorenzini 2020; Tarrow 2021.
15. Kitschelt 2006.
16. Della Porta, Fernández, et al. 2017, 9–10.
17. Diani 1992, 3.
18. The publications by these scholars in this formative period are among the five most cited publications in the literature on protest in both sociology and political science. Much of what followed can be explained as corrections and expansions (Barrie 2021, 924). McAdam (1982) 1999.
19. Olson 1965; Nash (2000) 2010, 99–100; Walder 2009, 394–398.
20. Snow et al. 1986; Jasper 2020.
21. Jasper 2018, 2020.
22. Tilly 1986, 7, 77, 384, 395.
23. Della Porta 2015, 5, 65. Della Porta draws from Guy Standing's pioneering work on the precariat.
24. Della Porta 2015, 23–24; Harvey 1989, part 2; Vachon et al. 2016, 3.
25. Kitschelt 2012, 148; Della Porta 2015, 4, 12.

CHAPTER 3. THE POLITICAL ENVIRONMENT AFTER THE GREAT RECESSION

1. This is typical of the role of digital media in social movements. Digital media amplify interest by helping people reveal their preferences to each other and discover common ground, but quick media mobilizations can dissipate without organizational efforts to create solidarities and bring people into meaningful action (Tufekci 2018, xxiii, 26). Movement media were generally produced by people participating in the movement networks in Brooklyn. The editors and reporters were part of the same leftist community, simultaneously inspiring the movement and trying to capture its ideas and sensibilities. The media texts didn't come from afar and were interpreted in their local environment. Movement participants frequently read individual articles based on recommendations from personal connections, but they were also not spending hours on social media platforms every day. Organizers in NYC-DSA gain much knowledge from embodied interactions with citizens and organizations in the city.
2. Harvey 1989, part 2; Dobratz et al. 2019, chap. 4; Agarwal et al. 2017. One of the lasting outcomes of the Great Recession is the expansion of homelessness. The culture of precarious workers living in vans is portrayed in the movie *Nomadland*, adopted from Jessica Bruder's namesake book of 2017. From the perspective of Michael Harrington (1962) and Barbara Ehrenreich (2001), we can understand homelessness not just in terms of rising costs of living in the city but also in terms of how society pushes poverty into obscurity.
3. Tascha Van Auken, interview with author, April 17, 2022.
4. Double Down News 2020; Cutlip 1994, 590, 624. Cutlip and others have identified the 1968 election as the moment when public relations took over politics (Cutlip 1994, 626). McNair (2004) has analyzed public frustrations with the public relations of the New Labour government in the UK, 1997–2004, similarly arguing that public relations has gained a more central function in politics and organized life more broadly. Edwards (2016) has attempted to strengthen the democratic potential of public relations.
5. Julia Salazar, interview with author, April 19, 2022; Sam Lewis, interview with author, May 1, 2022; William Rudebusch, interview with author, March 20, 2022.
6. Michael Paulson, interview with author, April 7, 2022.
7. Neal Meyer, interview with author, May 19, 2022.
8. Van Auken interview.
9. Van Auken interview.
10. Josh Kraushaar, interview with author, March 27, 2022. The experienced observer John Tarleton recognized this historical shift from reactive to proactive in political life. John Tarleton, interview with author, April 8, 2022.
11. Sanders 2016, 50, 52.
12. Sanders 2016, 46–47.

13 Day 2020, 5; *Jacobin* 2016, 17–18; Devon McManus, interview with author, April 1, 2022; Paulson interview; David Duhalde, interview with author, June 14, 2022; Kraushaar interview; Gustavo Gordillo, interview with author, May 10, 2022; Meyer interview.
14 John Tarleton, interview with author, April 8, 2022.
15 Van Auken interview.
16 Cohen 2015; Schneider and Kahn 2020. Sanders became a fierce advocate for rolling back neoliberal tax policies long before Occupy (Sanders with Gutman [1997] 2019, 245–246). He supported Bill Clinton's proposal for a wealth tax in 1993.
17 Nnaemeka and Luo 2021; New York State Senate 2000; Campanille 2020.
18 Glazer and Gillers 2021; Bellafante 2021; Karni 2021; Carras 2021. The opinions of *The New York Times* readership are registered in the many reader comments, especially on Karni 2021.
19 Kepple et al. 2022.
20 David Alexis, interview with author, May 8, 2022.
21 Tarleton interview.
22 Tefa Galvis, interview with author, April 8, 2022; Devon McManus, interview with author, April 1, 2022; Robert Wood, interview with author, March 11, 2022.
23 Writes Duhalde (2022), "People think that Bernie Sanders's campaign really drew people into the DSA. It wasn't. It's what got people aware of DSA, and then when Trump won, that's when thousands and thousands of people started joining."
24 Lewis interview. See the introduction.
25 Aina Lakha, interview with author, May 5, 2022.
26 This topic was widely discussed on TikTok after the murder of UnitedHealthcare CEO Brian Thompson. In January 2025, the controversial public speaker Scott Galloway exploited the moment by claiming that income inequality had reached a dangerous level. Galloway commented on MSNBC's *Morning Joe*, "We are in the midst of a series of small revolutions to correct income inequality. Income inequality is out of control. Our tax policy has gone full oligarch" (Boggioni 2025). He looked at the protest movements of the 2010s as indications of growing public frustration, but he did not acknowledge how they had addressed inequality and what solutions they had proposed. The viral video clip made it seem as if extreme inequality was a new phenomenon and that the 2016 Sanders presidential primary had not happened.

CHAPTER 4. ANTICOMMUNISM KILLS

1 Bousso 2023; Ravikumar and Twidale 2023; Carrington 2023; Noor 2023; McFarlane 2023.
2 Galbraith (1958) 1998, 189. For a survey of literature on how capitalism compromises the environment, see Holt 2022, 490–493.
3 Sanders studied at the University of Chicago. He acknowledges inspiration from Eugene Debs, a union and Socialist Party organizer in the late nineteenth and early twentieth centuries. He admires Debs's political practice. Sanders writes that

rising inequality motivated this decision to run in 2015: "How did I go from being 99 percent sure that I would not run for president in October 2013 to standing before a crowd of five thousand on May 26, 2015, in Burlington, Vermont, declaring my candidacy? . . . For me, the bottom line was that this country was facing enormous crises: the continued decline of the middle class, a grotesque level of income and wealth inequality, high rates of real unemployment, a disastrous trade policy, an inadequate educational system, and a collapsing infrastructure. On top of all that, we needed bold action to combat climate change" (2016, 52).

4 Nichols 2018.
5 Neil Meyer, interview with author, May 19, 2022; Newport 2018; Saad 2019.
6 Ronin Wood, interview with author, March 1, 2023.
7 Fukuyama (1992) 2006, 98.
8 Fukuyama (1992) 2006, 35, 291.
9 Brick 2013; Marcetic 2024.
10 Blair and Schröder 1998. See also Hall 1998.
11 Giddens 1998, 1, 4–5, 25.
12 Reed 2000, 128–129; Dreier 2010; Gitlin 2003, 108–109.
13 After the Russian Revolution in 1917, the federal government created a committee in 1918 to investigate the influence of the Bolsheviks in the United States. In the late 1940s, the government set up the House Un-American Activities Committee, which targeted leftists on the pretext that communism is a form of treason. The Red Scare also provided a rationale for the FBI to investigate and repress the leadership of the civil rights movement and the New Left in ways that many people felt were unconstitutional. During the anticommunist campaign at home in the 1950s, hundreds of people were imprisoned, ten thousand lost their jobs, and thousands were harassed or denied employment. This campaign was dwarfed by the country's international military campaign in subsequent decades, which led to the killing of millions of people in Vietnam, Indonesia, and Brazil (Schrecker 1998; Bevins 2020).
14 Avlon 2020; City University of New York, n.d.
15 H. Lewis 2021; Rebel Wisdom 2018. Prominent neoconservative figures in the 1980s include Irving Kristol and Allan Bloom.
16 Karpan 2022. In a TV interview on NY1 the day after Kristen Gonzalez's election in August 2022, the journalist Errol Louis asked her if socialism is not a radical ideology with demands for expropriation. Gonzalez, who had made no such claims, responded that her campaign had been targeted by "fear-mongering and divisive politics from Republicans" trying to scare the public about socialism. Her aim is to promote "a community of care" in the interest of "working class New Yorkers" (Garcia 2022).
17 In the video, the girl says, "I want to tell you about my plan to single-handedly save the planet. I call it the Green New Deal. I picked green because I'm still learning my colors." Fox's loyalty to Trump was particularly strong in its coverage of the Mueller investigation into Putin's election interference. Schatz 2019; Apen-

Sadler 2019; Markowicz 2018; Benner 2020; Benkler et al. 2018; Milman 2022; Lynch 2020; Place 2021; Julia Salazar, interview with author, April 19, 2022.
18 Tascha Van Auken, interview with author, April 17, 2022; Weiss 2018; Stephens 2018.
19 Rooney 2021, 18, 41.
20 Odell 2019, xiv.

CHAPTER 5. A DEPARTURE FROM THE DEMOCRATIC PARTY IN BROOKLYN

1 Collins 1986, 1999.
2 McAlevey 2020; Democratic Socialists of America, n.d.-c; Zimmerman 2022; Miller and Waggoner 2022.
3 Tim Hunter, interview with author, March 4, 2022; Hunter Rabinovitz, interview with author, March 15, 2022; Abraham Silverstein, interview with author, March 15, 2022.
4 Day and Uetricht 2020, x, xvii, xx, 153. There are no other book-length accounts by insiders with the same broad perspective on the organization. In a decision influenced by corporate donors, the party leadership maneuvered Sanders out of the 2020 primary in favor of Joe Biden.
5 Day and Uetricht 2020, 59, 80, 92. We learn about the rank-and-file program that was passed at the national convention in 2019. The authors report from members, such as Witthaus, who have participated in this program (Day and Uetricht 2020, 180, 186, 195).
6 Day and Uetricht 2020, 111, 141.
7 McAlevey 2020, 3-4. McAlevey argues that the Democratic Party has sold out to the economic elites and played along with the libertarianism of Silicon Valley tech corporations. She portrays Silicon Valley as a power bloc promoting right-wing, antiworker politics with Democratic-backed social positions, such as support for gay marriage, prochoice legislation, and ethnic diversity (3–7).
8 New York State Board of Elections, n.d. The national census does not provide information about the age group between eighteen and forty, but the following source offers a likely estimate: Neilsberg Research 2024.
9 Krase and LaCerra 1991, 28.
10 Krase and LaCerra 1991, 50, 131.
11 Krase and LaCerra 1991, 12, 22, 53, 65. The organized crime groups engaged in extortion, bribery, money laundering, gambling, and prostitution.
12 Krase and LaCerra 1991, 74, 125, 128, 131.
13 Giddens 1991; Rhoda Jacobs, interview with author, June 7, 2022.
14 Lakha 2021.
15 Joseph and Gonen 2022; Fink 2022; Rosenthal 2018; Bromwich and Rashbaum 2023; Rothfeld 2023; Louis 2022; Marsh and Hicks 2021. Adams was indicted on corruption charges in 2024. He was the first sitting mayor in the modern history of New York City to face criminal charge (Ransom 2024).

16 Grim 2022.
17 Rabinowitz 2022; Kawaguchi 2022; Democracy Now! 2022.
18 Moynihan and Glazer (1963) 1970.
19 Data USA, n.d. The community insider and political scientist François Pierre-Louis (2013) describes the area as follows: "On Nostrand, Church, and Flatbush Avenues in Brooklyn, there were numerous Haitian grocery stores, restaurants, dry cleaners, taxi stations that competed for immigrant businesses. There were three major Haitian weekly newspapers that published in Creole and French and one in English, aimed primarily at the children of the first-generation immigrants. In addition to newspapers that were written in French and Creole, the immigrants also opened several community centers to provide literacy and English courses. These centers also provided job referrals to newly arrived immigrants. These community centers were also places that people came to exchange news about the homeland, maintain contacts with friends and engage in political mobilization against the Duvalier regime."
20 I attended the annual Haitian Day Parade and Festival down Nostrand Avenue / Toussaint Louverture Boulevard in May 2022 and observed politicians and community leaders taking selfies while marching along with musicians and dancers. I canvassed regularly in four large apartment buildings in South Prospect Park at Argyle and Westminster Roads, where I talked to dozens of Haitian Americans who had lived in the neighborhood since the 1960s and in some cases even in the same apartment. On Sunday, April 24, 2022, around two o'clock in the afternoon, I was invited inside the home of a constituent named Pierre-Churchill Verneau, seventy-eight, on Argyle Road and had a long conversation with him and his wife about life in the neighborhood and about politics (Pierre-Churchill Verneau, conversation with author, April 24, 2022).
21 The demographics of the area west of Flatbush Avenue are 26 percent Black, 17.4 percent Hispanic, and 40 percent white, and the median income is $70,130. Bernie Sanders grew up in Midwood and remembers that the working class lived in rent-controlled apartments and the middle class in private houses. He lived on 1525 East Twenty-Sixth Street (Nir 2016). A low-income area near Ditmas Park includes the blocks between Cortelyou and Beverly Roads from Sixteenth Street to Flatbush Avenue (City-Data.com, n.d.-a; NYU Furman Center, n.d.-b; Mollenkopf 2023; Jacobs interview). East Flatbush is 78 percent Black, 4.1 percent Hispanic, and 4.1 percent white (City-Data.com, n.d.-b; NYU Furman Center, n.d.-a). Brooklyn Historical Society, n.d.
22 These buildings had mostly empty hallways, blank concrete and metal surfaces, and fluorescent lighting, and residents avoided eye contact and generally adopted self-isolating behaviors. There was drug dealing and smoking in the stairways, with a cold attitude, not at all relaxed.
23 Bichotte Hermelyn and Parker have long-standing ties with Yvette Clarke and her successors on the city council, Clarke's mother, Una Clarke, and Mathieu Eugene. The first local Haitians elected to public office were US Representa-

tive Yvette Clarke and city council member Mathieu Eugene in 2007, and Bichotte Hermelyn represented a further consolidation. Bichotte Hermelyn replaced the Jewish progressive Assembly Member Rhoda Jacobs in 2015 and became chair of the Brooklyn Democratic Party in 2020. The elected officials in Flatbush are Farah Louis and Rita Joseph (city council members), Bichotte Hermelyn (state assembly member and chair of the party's County Committee), Kevin Parker (state senator), and Yvette Clarke (US representative). Bichotte Hermelyn and Parker are the most powerful political leaders in Flatbush. Monrose is pastor of Mount Zion Church of God Seventh Day in East Flatbush and director of Faith-Based and Clergy Initiatives. He has been the director of a religious network at the Office of the Brooklyn Borough President since 2013.

24 David Alexis, interview with the author, May 8, 2022. Alexis cofounded the Drivers Cooperative, a rideshare company owned and operated by its employees.
25 Alexis interview.
26 Ballotpedia, n.d.
27 The Ecosocialist Working Group leaders were the ones driving the development of Alexis's campaign. They brought him into its fold, trained him, and helped him develop ties with the climate movement. Some recruiters said it was important for NYC-DSA to prove that it could run a working-class Black candidate in Flatbush. This picture of the situation is detailed further in the individual-level explorations in chapters 6–7. Daniel Goulden was part of the recruitment process. They remember, "I was the strategy cochair of the Ecosocialist Working Group and reached out to David [in 2020]. . . . I continued talking to David and was thinking, 'You know what, this is a guy living in Flatbush. He can organize with us.' I talked to him and to Grace [Mausser]. David was like, 'I'm a home health aide, and blackouts are a real health issue for a lot of my clients.' And normally, I have to explain our idea of public power, but he got it instantly. So we basically started integrating David into the Ecosocialist Working Group." Daniel Goulden, interview with author, May 12, 2022.
28 Jacobs interview. Jacobs and the progressive movement activists were opposed to the social injustice of low-income young Black people in the neighborhood being drafted in large numbers, while those who could afford to stay in college kept getting deferred. The redlining was supported by the machine, including the incumbent assemblyman and judges.
29 Kathryn Krase, interview with author, April 4, 2022. Krase followed her father's path of social engagement by gaining a social science PhD and working as a social worker and lawyer.
30 The progressive candidate was Jessica Ramos. For Kathryn Krase, DSA was a somewhat distant network with which she had little interaction. This was typical of Democrats over forty. DSA was off the radar for many simply because they were socialized into an older political culture and did not participate in younger

networks and media spaces. By contrast, most politically active leftists under forty were highly aware of DSA as a political identity among people their age. The progressives I interviewed generally started their day by reading *The New York Times*, and some were writing for the paper. They also read *The Atlantic*, *The Intercept*, and *The Guardian*. Kathryn had never read *Jacobin*.

CHAPTER 6. THE ALEXIS CAMPAIGNERS IN FLATBUSH

1 What is more, senate districts are larger than assembly districts, so the Alexis campaign needed to engage a broader constituency than NYC-DSA's assembly campaigns. The state senate district includes about 308,000 residents, whereas assembly districts generally include 129,000 residents (Ballotpedia, n.d.). The New York Assembly and Senate are the two houses of the bicameral state legislature.

2 "The state has more power than the city, and we're not at the level of organization where we can contest for power at the federal level in the same way that we can at the state level." Gustavo Gordillo, interview with author, May 10, 2022. New York State Division of the Budget 2023; New York State Senate, n.d.; Lane and Wolf 2012, 227, 235. There is widespread public skepticism about the state legislature in New York. The legislative process is dominated by three people: the governor, the senate majority leader, and the speaker of the assembly (Galie 2012, 40). Critics argue that members of both the senate and assembly spend too much of their time "running for reelection, serving as ombudsmen, and pursuing their private interests," rather than "occupying a prototypical legislative role" (Lane and Wolf 2012, 226).

3 Small and Calarco 2022, 24.

4 The meaning of the term "ecosocialism" in DSA will become clearer in chapter 9. For an early and more academic movement approach to the term, see Angus et al. 2008.

5 Alexis 2022a.

6 The Parker campaign made thirty-nine posts related to his campaign to Parker's Instagram account in the period January 2022 through August 2022. The Alexis campaign made 287 posts from October 2021 to September 2022, most of them related to his campaign.

7 Said the experienced DSA insider Gustavo Gordillo, "We have a lot of members in David's district and in Flatbush, but they're mostly transplants. There are not many of the Caribbean immigrants who live there or Black union members, like most of the electorate there." Gustavo Gordillo, interview with author, May 10, 2022.

8 Many trusted politicians, and many were interested in dialogue with politicians and had concrete suggestions for them. Stephen, an immigrant from Trinidad working as a janitor in a church, was frustrated that the police are "rude because they don't understand other cultures": "We have people from all over the world here. The police only understand 'American' culture." José, in his sixties, was concerned about gun violence and climate change, but he trusted politicians and felt that they were working for him.

9 Daniel Goulden and his wife, Nell Crumbley (creative writing and architecture, Cornell); Theresa Paquette (art history, NYU); Aina Lakha (comparative literature, Columbia); Rachel Himes (art history, Columbia); Andrew Butler (drama, NYU); Nadia Tykulsker (dance, University of Michigan); Robert Wood (music, City University); and Patrick Robbins (climate and society, Columbia University).

10 A candidate seeking endorsement from NYC-DSA said in a public meeting that the visuals are a major draw for him. "You have cool visuals!" Keron Alleyne, NYC-DSA candidate forum on Zoom, March 17, 2022.

11 Aina Lakha, interview with author, May 5, 2022; Gordillo 2024.

12 "I Am Culture" is a Brooklyn lifestyle brand promoting Black empowerment.

13 DSA for the Many is a kind of political action committee created in 2020. It collects and distributes funds to electoral campaigns in individual neighborhood branches. It also introduced the concept of running candidates across the city collectively as a slate on the same platform (Mellins 2020). Michael Pollak explains that Devon McManus was one of the people who invented DSA for the Many and that it is a solution to constraints in election law: "There are so many compliance rules designed to keep people out. A party can give unlimited amounts of money to a person, but we're not a party. The multicandidate committee solved the problem." Michael Pollak, interview with author, May 11, 2022. DSA for the Many is battling a lawsuit, however, that could have serious consequences for NYC-DSA. NYC-DSA claims to have followed the advice of the Board of Elections, but that the Division of Election Law Enforcement argued in 2023 that the committee ought to have filed additional paperwork, resulting in a fine of $300,000. NYC-DSA has challenged the imposition of the penalty and claims that the lawsuit is simply a malicious campaign against DSA run by the Hochul administration (NYC-DSA, email, January 27, 2025).

14 Daniel Goulden, interview with author, May 12, 2022.

15 Gordillo et al. 2022.

16 Devon has served on DSA's National Electoral Committee and has been both treasurer for NYC-DSA and Central Brooklyn representative on NYC-DSA's highest leadership body, the Steering Committee.

17 Devon McManus, interview with author, April 1, 2022.

18 McManus interview.

19 Robert Wood, interview with author, March 11, 2022.

20 Sarah Reibstein, interview with author, April 3, 2022.

21 Anderson 2005; Henninger 2005; Limbaugh 2009. Will studied mathematics and philosophy in college and taught mathematics in Sioux City, Iowa, for a few years before moving to Brooklyn.

22 William Rudebusch, interview with author, March 20, 2022.

23 Pollak interview; Muzzio 2012, 192.

24 Daniel Goulden, interview with author, May 10, 2022.

25 Josh Kraushaar, interview with author, March 27, 2022.

26 Ronin Wood, interview with author, March 1, 2023.

CHAPTER 7. ORGANIZATION AND LEADERSHIP IN NYC-DSA

1. For organizational diagrams of national and New York City DSA, see Democratic Socialists of America, n.d.-b; and NYC-DSA, n.d.-b.
2. Isserman 2000, chap. 11; Aronoff et al. 2020, 41–42; Fraser 1990.
3. J. Wilson 1962.
4. Meyer 2024.
5. Tolentino 2020; Powell 2020; Harrington 1962; Ehrenreich 2001.
6. Meyer 2016, 10. Meyer points out that the Green Party focuses on presidential campaigns and that it lacks resources for such campaigns.
7. Sam Lewis, email to author with a link to the full document, May 9, 2022.
8. New York State Division of the Budget 2023; Lane and Wolf 2012, 227, 231.
9. The Labor Branch is skeptical of the value of electoral work for base building, pointing to the temporary outlook of electoral campaigns and their dynamics of competition. It is also skeptical about the level of influence in state government, citing the small number of electeds and the highly centralized power structure in the legislature, including the majority legislative caucuses. In late 2023, the Labor Branch was folded into a new Labor Working Group that is trying to build on the success of the "Union Power" campaign and make organizing around labor issues open to all NYC-DSA members, not just rank-and-file existing union members. The Labor Working Group has a legislative subcommittee and more of a connection to electoral work.
10. Henwood 2019; Bread & Roses, n.d.; Red Star Caucus 2019. The other caucuses are: Anti-Zionist Slate, Groundwork, Constellation, Emerge, Communist Caucus, Libertarian Socialist Caucus, Marxist Unity Group, North Star, Red Labor, and Reform & Revolution.
11. Nohria and Khurana 2010.
12. Buhle and Wagner 2002.
13. Polletta 2002, 2020.
14. Skocpol 2003.
15. Sinnott and Gibbs 2014. Ganz helped organize the influential "Freedom Summer" delegation of the civil rights movement to the Democratic National Convention in 1964 and then organized agricultural workers for sixteen years. He drew from these experiences in developing the leadership approach that defined Obama's 2008 field operation, for which Ganz was a key organizing trainer and strategist (Ganz 2010; Sinnott and Gibbs 2014).
16. Hepp and Hasebrink 2018, 27–28; Jensen (2002) 2020, 5–6; Van Auken 2020.
17. Democratic Socialists of America, n.d.-a; NYC-DSA 2022 Convention 2022. One crisis evolved around the Palestine Solidarity Working Group's calls for the expulsion of Jamaal Bowman in 2021 (Barkan 2021a). Another crisis evolved from animosity against committee members in the National Discussion Board, resulting in committee members quitting.

18 Socialists in Office debrief meeting, 470 Vanderbilt Avenue, April 28, 2022. The $2,000 monthly stipend is in the 2023 budget (NPC, n.d.).
19 Van Auken 2020.
20 Tascha Van Auken, interview with author, April 17, 2022.
21 Aina Lakha, interview with author, May 5, 2022.
22 Van Auken interview.
23 Jack Gross, interview with author, May 13, 2022.
24 Julia Salazar, interview with author, April 19, 2022.
25 Gustavo Gordillo, interview with author, May 10, 2022.
26 Salazar interview.
27 Gordillo interview.
28 Tefa Galvis, interview with author, April 8, 2022.
29 Lakha interview.

CHAPTER 8. THE PARTY DIMENSION

1 Tascha Van Auken, interview with author, April 17, 2022.
2 Sam Lewis, interview with author, May 1, 2022.
3 Gustavo Gordillo, interview with author, May 10, 2022.
4 McIntosh 2021.
5 Van Auken interview.
6 Jasper 2018; Devon McManus, interview with author, April 1, 2022.
7 Van Auken interview.
8 Sam Lewis, email to author with a link to the full document, May 9, 2022.
9 Alt and Echola 2018; Alami et al. 2018. National DSA supports local campaigns by channeling funding and volunteers into the campaigns that it endorses. It has helped organize thousands of members from around the country into phone banking for New York candidates (Goldenberg et al. 2020; Porcelli and Max 2020).
10 Meyer 2016.
11 Ackerman 2016; Barkan 2021b. Ackerman also mentions that WFP was pressured to endorse Andrew Cuomo, for instance, after Cuomo threatened union leaders to choose between him and WFP.
12 Party loyalty is a major factor in the third-party discussion. Voters might have sympathy for a certain candidate but be steadfast on voting Democrat. Michael Pollak canvassed for Jabari Brisport's 2017 campaign and remembers how many people were unwilling to vote for a Green Party candidate. Brisport ran a DSA-backed campaign in 2019 and won. Pollak remembers, "That was a kind of a hilarious experience. On Election Day, I was doing 'get out the vote.' I was standing in front of this building, a huge housing project across the street from the polling place. So, you know, you can't go more than one hundred feet. I did this during the day because I work at night. One after another, all West Indian women, were coming out of this building, and they'd walk and smoke. So I had a nice, long time to talk to them. I had this card with this beautiful young man [Brisport] on

it, and they totally would have voted for this guy until I said he was a Green Party candidate." Michael Pollak, interview with author, May 11, 2022.
13 Meyer 2022; J.L. 2022.
14 Maisano 2022.
15 Maisano 2022.
16 David Duhalde, interview with author, June 6 and 14, 2022. The alienation felt by members who joined NYC-DSA in the 1980s is illustrated by the departure in late 2023 of two dozen such veteran members, including DSA cofounder Maurice Isserman. The veterans published farewell letters in *The Nation* and *The Atlantic* explaining that they were leaving because of the organization's "politically and morally corrupt" response to Hamas's massacre on October 7, 2023. It is clear from their letters that these veterans had little knowledge of NYC-DSA's "No Money for Massacres" campaign, which included one hundred thousand phone calls to Congress demanding a cease-fire. NYC-DSA was one of the first organizations to start the peace movement. Isserman criticized entryism, but millennials who had been in the organization for years reported that they had never seen the organization more united (Isserman 2023; Thier 2023; Powell 2023). Neal Meyer (2023) commented, "I've been a very active member in DSA for 11 years. I've never seen the org more united than it has been in the last 6 weeks. 24 disgruntled and totally inactive members quitting a 70,000+ member organization does not a crisis make, let's please have a sense of proportion!"
17 For a discussion of television and political campaign communication in the contemporary media system, see Chadwick 2017.

CHAPTER 9. THE ROAD TO GREEN NEW DEAL LEGISLATION

1 New York State Assembly 2023.
2 New York State Senate 2023; Allen 2024. The news about the Superfund was shared widely by climate protest movement networks on social media.
3 Uteuova 2023; Featherstone 2023a; Karolidis and Karcich 2023; Kang et al. 2023. The event obviously also got coverage in New York–based national leftist magazines, such as *Jacobin*, *The New Republic*, and *The Nation*.
4 Friedman 2007; Fisher 2024, 17, 44, 117; Pettifor 2019, 3–5. Schepelmann et al. 2009; Hawkins, n.d.; European Commission 2019; Huber 2022b.
5 Pettifor 2019, 13–14; N. Klein 2009; United Nations 2009.
6 I have decided to make this source anonymous here to protect them. The interview was conducted in spring 2024.
7 Atkin 2019; Greens / Green Party USA 2000. The Green Party's 2000 platform used the term "Global Green Deal" as part of the theme "International Solidarity." The platform was not framed around climate, and it has not made racial justice central to its Green New Deal concept. *Daily Free Press* 2012; Mufson 2012; Green Party, n.d.; Stein/Baraka 2016.
8 Public interest in climate was stimulated by documentaries of melting glaciers in the Arctic, books by popular intellectuals and journalists who had worked

on climate change for more than a decade, revelations of the fossil fuel industry's climate denial and lobbying, grave IPCC reports, and hurricanes in the Caribbean and the Gulf Coast (Holt 2022).
9 In 1988, Wirth was chair of the Energy and National Resource Committee that conducted the seminal hearing featuring the NASA scientist James Hansen. Garvey 2019; Cha 2022; Atkin 2022.
10 Gupta 2019, 31–33; Raskin 2022; Solnit 2019; Murphy 2018. The journalist Wendy Carrillo reported from the camp during those months and witnessed. The night of November 21 stands out in her memory: "I made my way to the front of the group, close to the barricade, my official Standing Rock press pass displayed outside my jacket. It was −20 degrees F. What I witnessed and experienced that night changed the course of my life. I saw people being hosed with water cannons—an international human rights violation given the frigid conditions. I saw people's clothes instantly freeze on their bodies. . . . Then, suddenly, canisters were flying over my head. The police were using tear gas. . . . Around me, I heard people throw up. I heard people cry out in pain. . . . The elders had been in a circle drumming, singing traditional Lakota songs. They, too, were teargassed. No one was spared. . . . I made it back to my tent that night, and woke up the following morning feeling like a different person. I had lost all fear. . . . It is clear to me now: The power of protest is real. The power of impact, of seeing firsthand people help strangers and build community in the fight for a greater good, of seeing what can and should be versus what is, is powerful, it's life altering" (Carrillo 2020, 138–141). The Standing Rock protests led President Obama to deny the permits in December. However, Trump signed an executive order expediting the pipeline construction in his first week in office (Lakhani 2022).
11 Aronoff 2018. Ocasio-Cortez has a minor in economics.
12 Mays 2019; Battistoni 2019; Yeampierre 2020. The Working Families Party (WFP) made a strong statement of support for AOC in 2019 when this organization moved left, aligning itself with the new millennial Left and wider anti-Trump movement and appointing thirty-one-year-old Sochie Nnaemeka as its new director.
13 Gupta 2019, 34; Brand New Congress 2017; Justice Democrats, n.d.; Smith 2021; Institute for Public Knowledge 2020.
14 US House of Representatives 2019; US Senate, n.d.
15 US House of Representatives 2019, 2; IPCC 2019, 31; Sen 1999, 3. The other report, *USGCRP 2018: Fourth National Climate Assessment*, was produced by federal government agencies. It analyzes the economic consequences of climate change broadly without an ethical agenda of equality and justice. The question of poverty appears in the analysis of adaptation capacity, but this is presented as a practical challenge rather than a central matter of moral concern (Reidmiller et al. 2018, 409). The first volume of *The Fourth National Climate Assessment* was published in 2017. It analyzes impacts on the physi-

cal Earth system. The second volume focuses on human welfare, societal, and environmental aspects.
16 US House of Representatives 2019, 4.
17 US House of Representatives 2019, 5.
18 US House of Representatives 2019, 13.
19 The IRA uses the term "underserved communities" instead of "vulnerable communities." The term is defined in the Senate proposal (US Senate 2022, 583). The IRA has a "Neighborhood Access and Equity Program" in the "Transportation and Infrastructure" part that seeks to "mitigate or remediate negative impacts on the human or natural environment resulting from a facility described in subsection (c)(2) in a disadvantaged or underserved community" (US House of Representatives 2022, section 60501).
20 Patrick Robbins, interview with author, March 29, 2024.
21 Featherstone 2023b.
22 N. Klein 2020; Democratic Socialists of America 2021. Klein spoke at a launch event for a national DSA Green New Deal campaign in 2021.
23 Gustavo Gordillo, interview with author, May 10, 2022.
24 New York Energy Democracy Alliance, n.d.
25 Ashford 2023. The article refers to the Climate Leadership and Protection Act of 2019.
26 Robbins interview.
27 Shay O'Reilly, interview with author, March 27, 2024.
28 O'Reilly interview.
29 Ecosocialist Working Group 2019a, 2019b.
30 O'Reilly interview.
31 Eisenberg et al. 2019; NYC-DSA 2021.
32 Robbins interview; Public Power NY Coalition, n.d.; O'Reilly interview. ESWG members Charlie Heller and Amber Ruther started working for the Alliance for a Green Economy after the coalition was created.
33 Bozuwa et al. 2021; Bowman et al. 2023.
34 Bozuwa et al. 2021. NYPA was created in 1931 as an alternative to the private utilities market but ended up serving only public entities. By 2021, NYPA served schools, health-care providers, public transportation in the New York City area, and eight hundred businesses.
35 Robbins interview.
36 Aronoff 2022.
37 Kennedy et al. 2023.
38 Daniel Goulden, interview with author, May 12, 2022.
39 Robbins interview.
40 Robbins interview.
41 Gordillo interview; Giambrone 2021.
42 The ESWG developed the campaigns of David Alexis, Illapa Sairitupac, Sarahana Shrestha, and Vanessa Agudelo. ESWG leader Andrea Guinn comanaged

the Kristen Gonzalez campaign. The ESWG was not involved in Samy Olivares's campaign in Bushwick. Gonzalez and Shrestha won.
43 Robbins interview.
44 Michael Paulson, interview with author, April 7, 2022.
45 Julianne Feaver, interview with author, April 17, 2024.
46 Michael Pollak, interview with author, May 11, 2022.
47 Feaver interview.
48 Feaver interview.

CONCLUSION 1

1 Howe 2024a, 2024b; Brown 2024.
2 Howard 2022; Siegel and Bram 2024; Freking 2023; Urban Institute 2024. The median monthly rent in New York City is $3,500.
3 Copernicus 2024; Carrington 2024; Abnett 2024; IHME 2021; Buchwald 2023; S. Wilson 2023; Fisher 2024, chap. 2.
4 A two-hundred-page report by the International Criminal Court described the war on Gaza as a war with genocidal intent (Patel and Imran 2024). Stockholm International Peace Research Institute 2024.

CONCLUSION 2

1 Winters 2011; Mayer 2016; Massoglia and Evers-Hillstrom 2021; Ocampo 2022.

REFERENCES

Abnett, K. 2024. "World's Record-Breaking Temperature Streak Extends Through April." *Reuters*, May 8. www.reuters.com.

Ackerman, S. 2016. "A Blueprint for New Party." *Jacobin* 23 (Fall): 102–111.

Adam, D., and J. Randerson. 2010. "Secret Copenhagen Recordings Reveals Resistance from China and India." *The Guardian*, May 7. www.theguardian.com.

Agarwal, S., G. Amromin, I. Ben-David, S. Chomsisengphet, T. Piskorski, and A. Seru. 2017. "Policy Intervention in Debt Renegotiation: Evidence from the Home Affordable Modification Program." *Journal of Political Economy* 125 (3): 654–712.

Alami, A., T. Alt, Z. Echola, and N. Midir. 2018. "Democratic Socialists of America: National Political Committee: Minutes of Meeting of January 26–28, 2018." Democratic Socialists of America, March 22. www.dsausa.org.

Alexis, David. 2022a. "Thank You!" Campaign email, August 26.

Alford, R., and R. Friedland. 1985. *Powers of Theory: Capitalism, the State and Democracy*. New York: Cambridge University Press.

Allen, J. 2024. "New York to Fine Fossil Fuel Companies $75 Billion Under New Climate Law." *Reuters*, December 26. www.reuters.com.

Alt, T., and Z. Echola. 2018. "Democratic Socialists of America: National Convention August 4–6, 2017, University of Illinois at Chicago, Chicago IL." Democratic Socialists of America, July 19. www.dsausa.org.

Amelang, S. 2024. "'No Prospects for Success'—Austrian Branch of Climate Activist Group Last Generation Disbands." *Clean Energy Wire*, August 7. www.cleanenergywire.org.

Anderson, B. 2005. *South Park Conservatives: The Revolt Against Liberal Media Bias*. New York: Regnery.

Angus, I., J. Kovel, and M. Löwy. 2008. "The Belem Ecosocialist Declaration." *Climate & Capitalism*, December 16. https://climateandcapitalism.com.

Apen-Sadler, D. 2019. "'I Call It the Green New Deal Because I'm Still Learning My Colors': Eight-Year-Old AOC Impersonator Mocks the Congresswoman's Climate Change Plan Saying It Was Based on *Ice Age: The Meltdown*." *Daily Mail*, May 16. www.dailymail.co.uk.

Aronczyk, M., and M. Espinoza. 2022. *A Strategic Nature: Public Relations and the Politics of American Environmentalism*. Oxford: Oxford University Press.

Aronoff, K. 2018. "Alexandria Ocasio-Cortez on Why She Wants to Abolish ICE and Upend the Democratic Party." *In These Times*, June 25. https://inthesetimes.com.

Aronoff, K. 2022. "Why Did Democrats and Solar Companies Just Kill a Climate Bill in New York?" *New Republic*, June 8. https://newrepublic.com.

Aronoff, K., P. Dreier, and M. Kazin, eds. 2020. *We Own the Future: Democratic Socialism—American Style*. New York: New Press.

Ashford, G. 2023. "In Rare Show of Force, House Democrats Pressure Hochul on Climate Bill." *New York Times*, March 29. www.nytimes.com.

Associated Press. 2023. "Thunberg and Nakate Slam Corporate World." *AP Newsroom*, January 19. https://newsroom.ap.org.

Atkin, E. 2019. "The Democrats Stole the Green Party's Best Idea." *New Republic*, February 22. https://newrepublic.com.

Atkin, E. 2022. "AOC Is HEATED." Substack page of the newsletter HEATED, September 16. https://heated.world.

Avlon, J. 2020. "Trumps Drops 'Socialism' Tack and Goes After 'Defund the Police.'" *CNN*, June 10. https://edition.cnn.com.

Ballotpedia. n.d. "New York State Senate District 21." Accessed September 14, 2023. https://ballotpedia.org.

Banerjee, N., J. Cushman, Jr., D. Hasemyer, and L. Song. 2015. *Exxon: The Road Not Taken*. InsideClimate News. https://vdoc.pub.

Barkan, R. 2021a. "Purge at DSA: Why Are Activists Trying to Expel Representative Bowman?" *The Nation*, November 22. www.thenation.com.

Barkan, R. 2021b. "What Is the Working Families Party?" *Political Currents by Ross Barkan*, April 19. https://rosselliotbarkan.com.

Barrett, P. 2009. "Rational Irrationality." *New York Times*, November 12. www.nytimes.com.

Barrie, C. 2021. "Political Sociology in a Time of Protest." *Current Sociology* 69 (6): 919–942.

Battistoni, A. 2019. "'Every Climate Deadline Is Important': An Interview with Alyssa Battistoni." *Socialist Forum*, Winter. https://socialistforum.dsausa.org.

Beck, U., and E. Beck-Gernsheim. 2002. *Individualization: Institutionalized Individualism and Its Social and Political Consequences*. London: Sage.

Bellafante, G. 2021. "New York, Finally, Taxes the Rich." *New York Times*, April 9. www.nytimes.com.

Benjamin, G. 2012. "The Study of New York Government." In *The Oxford Handbook of New York State Government and Politics*, edited by G. Benjamin. Oxford: Oxford University Press.

Benkler, Y., R. Faris, and H. Roberts. 2018. *Network Propaganda: Manipulation, Disinformation, and Radicalization in American Politics*. New York: Oxford University Press.

Benner, K. 2020. "Why the Mueller Investigation Failed." *New York Times*, August 5. www.nytimes.com.

Berry, J., and S. Sobieraj. 2014. *The Outrage Industry*. Oxford: Oxford University Press.

Bevins, V. 2020. *The Jakarta Method: Washington's Anticommunist Crusade and the Mass Murder Program That Shaped Our World*. New York: Public Affairs.

Blair, T., and G. Schröder. 1998. "Europe: The Third Way/Die Neue Mitte." Friedrich Ebert Foundation, June. https://library.fes.de.

Boggioni, T. 2025. "'CEO's Being Murdered in the Street': Expert Warns U.S. Headed Towards 'Small Revolutions.'" *RawStory*, January 8. www.rawstory.com.
Bottomore, T. 1979. *Political Sociology*. London: Hutchinson.
Bourdieu, P. 1990. "Lecture of 18 January 1990." In *On the State: Lectures at Collège de France 1989–1992*, 3–22. Cambridge, UK: Polity.
Bourdieu, P. 1996. *The State Nobility: Elite Schools in the Field of Power*. Cambridge: Cambridge University Press.
Bourdieu, P. (2002) 2008. *Political Interventions: Social Science and Political Action*. London: Verso.
Bousso, R. 2023. "Big Oil Doubles Profits in Blockbuster 2022." *Reuters*, February 8. www.reuters.com.
Bowman, J., A. Ocasio-Cortez, N. Velázquez, Y. Clarke, J. Nadler, A. Espaillat, and D. Goldman. 2023. Letter to Governor Kathy Hochul. *New York Times Documentation*, March 29. https://int.nyt.com.
Bozuwa, J., T. Riofrancos, S. Knuth, P. Robbins, S. Baker, A. McCullough, K. McDonald, C. Mackin, D. Cohen, B. Fleming, N. Graetz, and N. Shah. 2021. *A New Era of Public Power: A Vision for New York Power Authority in Pursuit of Climate Justice*. The Climate + Community Project, April 22. www.climateandcommunity.org.
Brand New Congress. 2017. "Platform." June. https://brandnewcongress.org.
Bread & Roses. n.d. "Where We Stand." Accessed October 6, 2023. https://breadandrosesdsa.org.
Brick, H. 2013. "The End of Ideology Thesis." In *The Oxford Handbook of Political Ideologies*, edited by M. Freeden, L. Sargent, and M. Stears. Oxford: Oxford University Press.
Bromwich, J., and W. Rashbaum. 2023. "District Attorney's Investigations Burrow into Adams's Circle of Support." *New York Times*, August 3. www.nytimes.com.
Brooklyn Historical Society. n.d. "Slavery in Brooklyn." An American Family Grows in Brooklyn: The Lefferts Family Papers. Accessed September 18, 2023. https://lefferts.brooklynhistory.org.
Brown, A. 2013. "Communism." In *The Oxford Handbook of Political Ideologies*, edited by M. Freeden, L. Sargent, and M. Stears. Oxford: Oxford University Press.
Brown, H. 2024. "Why AOC's Case for Impeaching Thomas and Alito Is Strong." *MSNBC*, July 13. www.msnbc.com.
Brown, W. 2015. *Undoing the Demos: Neoliberalism's Stealth Revolution*. New York: Zone Books.
Brown, W., P. Gordon, and M. Pensky. 2018. *Authoritarianism: Three Inquiries in Critical Theory*. Chicago: University of Chicago Press.
Bruder, J. 2017. *Nomadland: Surviving America in the Twenty-First Century*. New York: Norton.
Brulle, R., and R. Dunlap. 2015. "Sociology and Global Climate Change." In *Climate Change and Society: Sociological Perspectives*, edited by R. Dunlap and R. Brulle. New York: Oxford University Press.
Bryan, K., C. Hogson, and J. Tauschinski. 2024. "Big Tech's Bid to Rewrite the Rules on Net Zero." *Financial Times*, August 14. www.ft.com.

Bryan, K., and A. Mooney. 2024. "How Companies Are Starting to Back Away from Green Targets." *Financial Times*, June 21. www.ft.com.

Buchwald, E. 2023. "Climate Change Is Costing the US $150 Billion a Year. Here's What That Looks Like." *CNN*, November 30. https://edition.cnn.com.

Buhle, P., and D. Wagner. 2002. "The Left and Popular Culture: Film and Television." *Monthly Review* 54 (3): 43–58.

Buller, A. 2022. "What's Really Behind the Failure of Green Capitalism?" *The Guardian*, July 26. www.theguardian.com.

Campanille, C. 2020. "AOC's Democratic Socialists Call NYC Voters to Back Tax Hikes on Rich." *New York Post*, December 8. https://nypost.com.

Carras, C. 2021. "AOC Defends Polarizing 'Tax the Rich' Met Gala Dress: 'The Medium Is the Message.'" *Los Angeles Times*, September 14. www.latimes.com.

Carrillo, W. 2020. "What AOC and I Learned at Standing Rock." In *AOC: The Fearless Rise and Powerful Resonance of Alexandria Ocasio-Cortez*, edited by L. Lopez. New York: St. Martin's.

Carrington, D. 2023. "'Insanity': Petrostates Planning Huge Expansion of Fossil Fuels, Says UN Report." *The Guardian*, November 8. www.theguardian.com.

Carrington, D. 2024. "We Asked 380 Climate Scientists What They Felt About the Future." *The Guardian*, May 8. www.theguardian.com.

Central Brooklyn Branch of the New York City Chapter of the Democratic Socialists of America. 2024. "An Office Grows in Brooklyn: A Proposal to Rent 320 Tompkins Ave." July. https://docs.google.com.

Cha, J. 2022. Testimony of Dr. J. Mijin Cha, Associate Professor, Occidental College Fellow, Worker Institute, Cornell University. In *Hearing on Fueling the Climate Crisis: Examining Big Oil's Prices, Profits, and Pledges*. House Committee on Oversight and Reform, United States House of Representatives, September 15. http://docs.house.gov.

Chadwick, A. 2017. *The Hybrid Media System: Politics and Power*. 2nd ed. Oxford: Oxford University Press.

Chakrabarty, D. 2009. "The Climate of History: Four Theses." *Critical Theory* 35 (Winter): 197–222.

Chibber, V. 2022. *The Class Matrix: Social Theory After the Cultural Turn*. Cambridge, MA: Harvard University Press.

Chomsky, N. 1987. *On Power and Ideology: The Managua Lectures*. Chicago: Haymarket Books.

Chomsky, N. 2002. "Democracy Under Capitalism." In *Understanding Power: The Indispensable Guide to Chomsky*. London: Vintage Books.

City-Data.com. n.d.-a. "Ditmas Park Neighborhood in Brooklyn, New York (NY), 11226 Detailed Profile." Accessed September 14, 2023. www.city-data.com.

City-Data.com. n.d.-b. "East Flatbush Neighborhood in Brooklyn, New York (NY), 11236, 11203, 11212 Detailed Profile." Accessed September 15, 2023. www.city-data.com.

City University of New York. n.d. "NYC Election Atlas." Accessed February 6, 2025. www.electionatlas.nyc.

Climate Litigation Network. n.d. "Our Story." Accessed August 13, 2024. https://climatelitigationnetwork.org.

Cohen, P. 2015. "What Could Raising Taxes on the 1% Do?" *New York Times*, October 16. www.nytimes.com.

Collins, P. 1986. "Learning from the Outsider Within: The Sociological Significance of Black Feminist Thought." *Social Problems* 33 (6): S14–S32.

Collins, P. 1999. "Reflections on the Outsider Within." *Journal of Career Development* 26 (1): 85–88.

Coombs, T., and E. Tachkova. 2024. "How Emotions Can Enhance Crisis Communication: Theorizing Around Moral Outrage." *Journal of Public Relations Research* 36 (1): 6–22.

Copernicus. 2024. "June 2024 Marks 12th Month of Global Temperatures at 1.5°C Above Pre-Industrial Levels." July 10. https://climate.copernicus.eu.

Coser, L. 1966. *The Functions of Social Conflict*. Glencoe, IL: Free Press.

Cutlip, S. 1994. *The Unseen Power: Public Relations: A History*. Hillsdale, NJ: Lawrence Erlbaum.

Daily Free Press. 2012. "Stein Promotes Green New Deal in Boston Tour Stop." October 11. https://dailyfreepress.com.

Data USA. n.d. "Flatbush & Midwood PUMA, NY." Accessed February 14, 2023. https://datausa.io.

Day, M. 2020. "We Won't Forget the Questions Bernie Asked." *Jacobin* 38 (Summer): 4–6.

Day, M., and M. Uetricht. 2020. *Bigger than Bernie: How We Can Win Democratic Socialism in Our Time*. New updated ed. London: Verso.

de Schweinitz, R. 2015. "'The Proper Age for Suffrage': Vote 18 and the Politics of Age from World War II to the Age of Aquarius." In *Age in America: The Colonial Era to the Present*, edited by C. Field and N. Syrett. New York: New York University Press.

della Porta, D. 2015. *Social Movements in Times of Austerity*. Cambridge, UK: Polity.

della Porta, D. 2020. *How Social Movements Can Save Democracy: Democratic Innovations from Below*. Cambridge, UK: Polity.

della Porta, D., J. Fernández, H. Kouki, and L. Mosca. 2017. *Movement Parties Against Austerity*. Cambridge, UK: Polity.

della Porta, D., F. O'Connor, M. Portos, and A. Ribas. 2017. *Social Movements and Referendums from Below: Direct Democracy in the Neoliberal Crisis*. Bristol, UK: Policy.

Democracy Now! 2022. "Democrats May Lose U.S. House Because New York Dem. Leaders Were Too Focused on Defeating the Left." November 10. www.democracynow.org.

Democratic Socialists of America. 2021. "Workers & the World, Unite! DSA for PRO Act Campaign Launch Call w/ Naomi Klein, Sara Nelson & More." YouTube, March 7. www.youtube.com.

Democratic Socialists of America. n.d.-a. "Constitution." Accessed December 27, 2022. www.dsausa.org.

Democratic Socialists of America. n.d.-b. "Leadership and Structure." Accessed December 27, 2022. www.dsausa.org.

Democratic Socialists of America. n.d.-c. "YDSA and RFS." Accessed September 26, 2022. https://rfs.dsausa.org.

Deutsche Welle. 2024. "COP29: World Leaders Meet in Baku, with Big Names Missing." November 12. www.dw.com.

Diani, M. 1992. "The Concept of Social Movement." *Sociological Review* 40 (1): 1–25.

Dobratz, B., L. Waldner, and T. Buzzell. 2019. *Power, Politics, and Society: An Introduction to Political Sociology.* 2nd ed. New York: Routledge.

Double Down News. 2020. "David Graeber on the Extreme 'Centre.'" YouTube, October 12. www.youtube.com/watch?v=-9afwZON8dU.

Dreier, P. 2010. "The ACORN Conspiracy, Continued." *American Prospect*, March 23. https://prospect.org.

Dreyer, J., and P. Locker. 2021. "Who's Who in DSA: A Guide to DSA Caucuses." Reform & Revolution, September 8. https://reformandrevolution.org.

Duhalde, D. 2020. "It's Party Time: As Socialists, We Need to Help Decide Who Runs the Democratic Party." *Dissent*, Winter, www.dissentmagazine.org.

Duhalde, D. 2021. "Socialists Across Generations: We Need to Talk." *Democratic Left*, Spring. https://democraticleft.dsausa.org.

Duhalde, D. 2022. "The State of American Socialism." *Socialist Forum*, Summer. https://socialistforum.dsausa.org.

Dyrhauge, H. 2020. "Political Myths in Climate Leadership: The Case of Danish Climate and Energy Pioneership." *Scandinavian Political Studies* 44 (1): 13–33.

Economist Intelligence Unit. 2023. *Democracy Index 2023: Age of Conflict.* www.eiu.com.

Ecosocialist Working Group, NYC-DSA. 2019a. "Campaign Overview for Public Power." Unpublished proposal, May 11.

Ecosocialist Working Group, NYC-DSA. 2019b. "Energy Rights Campaign." Unpublished proposal, January 9.

Edwards, L. 2016. "The Role of Public Relations in Deliberative Systems." *Journal of Communication* 66:60–81.

Ehrenreich, B. 2001. *Nickel and Dimed: On (Not) Getting By in America.* New York: Holt.

Eisenberg, A., A. Guinn, G. Gordillo, J. Neimeister, M. Paulson, and N. Shimpi. 2019. "Citywide Priority Campaign for a Green New Deal." Ecosocialist Working Group, New York City Chapter of the Democratic Socialists of America. Unpublished proposal, September 9.

Elstub, S., and Z. Khoban. 2023. "Citizens' Assemblies: A Critical Perspective." In *De Gruyter Handbook of Citizens' Assemblies*, edited by M. Reuchamps, J. Vrydagh, and Y. Welp. Berlin: De Gruyter.

European Commission. 2019. "The European Green Deal: Communication from the Commission to the European Parliament, the European Council, the Council, the

European Economic and Social Committee and the Committee of the Regions." December 11. https://eur-lex.europa.eu.

Extinction Rebellion UK. 2023. "Extinction Rebellion's New Year's Resolution: WE QUIT." January 1. https://extinctionrebellion.uk.

Extinction Rebellion UK. 2024. "2024: Now We Step It Up." January 1. https://extinctionrebellion.uk.

Falzon, D., J. Roberts, and R. Brulle. 2021. "Sociology and Climate Change: A Review and Research Agenda." In *Handbook of Environmental Sociology*, edited by B. Caniglia, A. Jorgenson, S. Malin, L. Peek, D. Pellow, and X. Huang. Cham, Switzerland: Springer.

Featherstone, L. 2023a. "Climate Activists Transform New York City." *Ethical Hour*, October 17. https://ethicalhour.com.

Featherstone, L. 2023b. "New York Socialists Won Big on Climate: How Did It Happen?" *In These Times*, September 12. https://inthesetimes.com.

Fink, Z. 2022. "Brooklyn Democratic Party Leader's Future in Doubt." Spectrum News NY1, July 11. https://ny1.com.

Fisher, D. 2006. *Activism, Inc.: How the Outsourcing of Grassroots Campaigns Is Strangling Progressive Politics in America*. Stanford: Stanford University Press.

Fisher, D. 2019. *American Resistance: From the Women's March to the Blue Wave*. New York: Columbia University Press.

Fisher, D. 2024. *Saving Ourselves: From Climate Shocks to Climate Action*. New York: Columbia University Press.

Fletcher, C., W. Ripple, T. Newsome, P. Barnard, K. Beamer, A. Behl, J. Bowen, M. Cooney, E. Crist, C. Field, K. Hiser, D. Karl, D. King, M. Mann, D. McGregor, C. Mora, N. Oreskes, and M. Wilson. 2024. "Earth at Risk: An Urgent Call to End the Age of Destruction and Forge a Just and Sustainable Future." *PNAS Nexus* 3:1–20.

Fraser, N. 1990. "Rethinking the Public Sphere: A Contribution to the Critique of Actually Existing Democracy." *Social Text* 25/26 (1990): 56–80.

Fraser, N. 2023. *Cannibal Capitalism: How Our System Is Devouring Democracy, Care, and the Planet—and What We Can Do About It*. London: Verso.

Freeman, J. 2000. *Working-Class New York: Life and Labor Since World War II*. New York: Free Press.

Freking, K. 2023. "US Homelessness Up 12% to Highest Reported Level as Rents Soar and Coronavirus Pandemic Aid Lapses." *Associated Press*, December 16. https://apnews.com.

Friedman, T. 2007. "The Power of Green." *New York Times*, April 15, www.nytimes.com.

Fukuyama, F. (1992) 2006. *The End of History and the Last Man*. New York: Free Press.

Galbraith, J. (1958) 1998. *The Affluent Society*. 40th anniversary ed. New York: Penguin.

Galie, P. 2012. "The New York Constitution and the Federal System." In *The Oxford Handbook of New York State Government and Politics*, edited by G. Benjamin. Oxford: Oxford University Press.

Ganz, M. 2010. "Leading Change: Leadership, Organization, Social Movements." In *The Handbook of Leadership Theory and Practice*, edited by N. Nohria and R. Khurana. Danvers, MA: Harvard Business School Press.

García, C., P. Duncan, L. O'Carroll, and J. Pinster. 2024. "Quarter of Political Donations in Europe Are Going to Extremist and Populist Parties, Data Reveals." *Follow the Money*, May 30. www.ftm.eu.

Garcia, D. 2022. "27-Year-Old Democratic Socialist Details Her Primary Win." *NY1*, August 26. www.ny1.com.

Garnham, N. 1979. "Contribution to a Political Economy of Mass-Communication." *Media, Culture, and Society* 1:123–146.

Garvey, E. 2019. Written statement to the Congressional Oversight Committee. In *Hearing on Examining the Oil Industry's Efforts to Suppress the Truth about Climate Change*, House Committee on Oversight and Reform, United States House of Representatives, October 23. http://docs.house.gov.

Gay, P. du, and G. Morgan. 2013. "Understanding Capitalism: Crises, Legitimacy, and Change Through the Prism of the New Spirit of Capitalism." In *New Spirits of Capitalism? Crises, Justifications, and Dynamics*, edited by P. du Gay and G. Morgan. New York: Oxford University Press.

Gessen, M. 2020. *Surviving Autocracy*. New York: Riverhead Books.

Giambrone. 2021. "'At the End of the Day, Climate Is a Working Class Issue': An Interview with Sarahana Shrestha." *Jacobin*, August 11. https://jacobin.com.

Giddens, A. 1991. *Modernity and Self-Identity: Self and Society in the Late Modern Age*. Cambridge, UK: Polity.

Giddens, A. 1998. *The Third Way: The Renewal of Social Democracy*. Cambridge, UK: Polity.

Gitlin, T. 2003. *Letters to a Young Activist*. New York: Basic Books.

Giugni, M., and J. Lorenzini. 2020. "The Politics of Economic Crisis: From Voter Retreat to the Rise of New Populisms." In *The New Handbook of Political Sociology*, edited by T. Janoski, C. De Leon, J. Misra, and I. Martin. Cambridge: Cambridge University Press.

Glazer, E., and H. Gillers. 2021. "New York City's Wealthy Will Pay Nation's Highest Tax Rates: How Will That Affect a Rebound?" *Wall Street Journal*, April 8. www.wsj.com.

Goldberg, M., X. Wang, J. Marlon, J. Carman, K. Lacroix, J. Kotcher, S. Rosenthal, E. Maibach, and A. Leiserowitz. 2021. *Segmenting the Climate Change Alarmed: Active, Willing, and Inactive*. New Haven, CT: Yale Program on Climate Change Communication. https://climatecommunication.yale.edu.

Goldenberg, S., J. Custodio, and J. Anuta. 2020. "As Their Reach Grows in Albany, Democratic Socialists Target the City Council." *Politico*, November 13. www.politico.com.

Goodwin, J., and J. Jasper, eds. 2004. *Rethinking Social Movements: Structure, Meaning, and Emotion*. Lanham, MD: Rowman and Littlefield.

Gordillo, G. (@unionGustavo). 2024. "The majority of new members to join @nycDSA since October 7th are people of color." X, May 30. https://x.com/unionGustavo/status/1795978088871768250.

Gordillo, G., N. Tykulsker, and I. Sairitupac. 2022. "We Went to Court for Climate, but Albany Is the Real Culprit." *The Indypendent*, January 20. https://indypendent.org.

Graeber, D., and D. Wengrow. 2021. *The Dawn of Everything: A New History of Humanity*. New York: Penguin.

Green Party. n.d. "Green New Deal." Accessed January 18, 2023. www.gp.org.

Greens / Green Party USA. 2000. "Platform of the Greens / US Green Party." www.greenparty.org.

Grim, R. 2022. "AOC: The New York State Democratic Party's Corruption May Have Cost Democrats the House." *The Intercept*, November 10. https://theintercept.com.

Guardian. 2022. "Greta Thunberg to Skip 'Greenwashing' Cop27 Climate Summit in Egypt." October 31. www.theguardian.com.

Gupta, P. 2019. *AOC: Fighter, Phenom, Change Maker*. New York: Workman.

Habermas, J. 1989a. *The New Conservatism*. Cambridge, UK: Polity.

Habermas, J. 1989b. *The Philosophical Discourse of Modernity*. Cambridge, UK: Polity.

Habermas, J. 1996. *Between Facts and Norms: Contributions to a Discourse Theory of Law and Democracy*. Cambridge, UK: Polity.

Hacker, J., and P. Pierson. 2015. "Confronting Asymmetric Polarization." In *Solutions to Political Polarization in America*, edited by N. Persily. Cambridge: Cambridge University Press.

Hall, S. 1998. "The Great Moving Nowhere Show." *Marxism Today*, November–December, 9–14.

Harrington, M. 1962. *The Other America*. New York: Macmillan.

Harvey, D. 1989. *The Condition of Postmodernity: An Enquiry into the Origins of Cultural Change*. Cambridge, MA: Blackwell.

Harvey, D. 2005. *A Brief History of Neoliberalism*. Oxford: Oxford University Press.

Hattenstone, S. 2021. "The Transformation of Great Thunberg." *The Guardian*, September 25. www.theguardian.com.

Hawkins, H. n.d. "Origins of the Green New Deal Slogan." Howie Hawkins for Our Future website. Accessed January 30, 2023. https://howiehawkins.us.

Henninger, D. 2005. "Rush to Victory." *Wall Street Journal*, April 29. www.opinionjournal.com.

Henwood, D. 2019. "The Socialist Network: Inside DSA's Struggle to Move into the Political Mainstream." *New Republic*, May 16. https://newrepublic.com.

Hepp, A., and U. Hasebrink. 2018. "Researching Transforming Communications in Times of Deep Mediatization: A Figurational Approach." In *Communicative Figurations: Transforming Communications in Times of Deep Mediatization*, edited by A. Hepp, A. Breiter, and U. Hasebrink. Cham, Switzerland: Palgrave Macmillan.

Hernandez, K., and B. Huang. 2023. "Growth and Development Committee 2023: Report to the DSA National Convention." Google Drive, https://drive.google.com.

Hersh, E. 2020. *Politics Is for Power: How to Move Beyond Political Hobbyism, Take Action, and Make Real Change*. New York: Scribner.

Hildyard, K., and D. Wolfe. 2002. "Child Neglect: Developmental Issues and Outcomes." *Child Abuse and Neglect* 26:679–695.

Hobsbawm, E. 2011 *How to Change the World: Reflections on Marx and Marxism*. New Haven, CT: Yale University Press.

Holt, F. 2022. "The Geo-Ecological Turn in Sociology and Its Implications for the Sociology of the Arts." *Cultural Sociology* 16 (4): 486–502.

Howard, R. 2022. "Working for a Living Wage." University of North Carolina at Chapel Hill, School of Government ncIMPACT Initiative, July. https://ncimpact.sog.unc.edu.

Howe, A. 2024a. "Justices Rule Trump Has Some Immunity from Prosecution." *SCOTUSblog*, July 1. www.scotusblog.com.

Howe, A. 2024b. "Supreme Court Strikes Down Chevron, Curtailing Power of Federal Agencies." *SCOTUSblog*, June 28. www.scotusblog.com.

Huber, M. 2022a. *Climate Change as Class War*. London: Verso.

Huber, M. 2022b. "How the Green New Deal Became the Inflation Reduction Act and Lost Its Soul." *Socialist Call*, August 17. https://socialistcall.com.

IHME (Institute for Health Metrics and Evaluation, University of Washington). 2021. "The Lancet: Extreme Heat Is a Clear and Growing Health Issue, with Evidence-Based Adaptation Plans Urgently Needed to Prevent Unnecessary Deaths." August 19. www.healthdata.org.

Institute for Public Knowledge. 2020. "Discussion: The Green New Deal." Panel discussion, April 15, 2019, featuring Rhiana Gunn-Wright. YouTube, August 10. www.youtube.com/watch?v=fHc1bkmi21E.

IPCC (Intergovernmental Panel on Climate Change). 2019. *IPCC, 2018: Global Warming of 1.5°C. An IPCC Special Report on the Impacts of Global Warming of 1.5°C Above Pre-industrial Levels and Related Global Greenhouse Gas Emission Pathways, in the Context of Strengthening the Global Response to the Threat of Climate Change, Sustainable Development, and Efforts to Eradicate Poverty*. Edited by V. Masson-Delmotte, P. Zhai, H.-O. Pörtner, D. Roberts, J. Skea, P.R. Shukla, A. Pirani, W. Moufouma-Okia, C. Péan, R. Pidcock, S. Connors, J. B. R. Matthews, Y. Chen, X. Zhou, M. I. Gomis, E. Lonnoy, T. Maycock, M. Tignor, and T. Waterfield. Geneva: IPCC. www.ipcc.ch.

Isserman, M. 2000. *The Other American: The Life of Michael Harrington*. New York: PublicAffairs.

Isserman, M. 2023. "Why I Just Quit DSA." *The Nation*, October 23. www.thenation.com.

Jackson, B. 2013. "Social Democracy and Democratic Socialism." In *The Oxford Handbook of Political Ideologies*, edited by M. Freeden, L. Sargent, and M. Stears. Oxford: Oxford University Press.

Jacobin. 2016. "What Did Bernie Do? A Conversation with Cedric Johnson, Matt Karp, and Jennifer Roesch." Volume 23 (Fall): 17–23.

Janoski, T., C. De Leon, J. Misra, I. Martin. 2020. "Introduction: New Directions in Political Sociology." *The New Handbook of Political Sociology*, edited by T. Janoski, C. De Leon, J. Misra, and I. Martin. Cambridge: Cambridge University Press.

Jasper, J. 2018. *The Emotions of Protest*. Chicago: University of Chicago Press.

Jasper, J. 2020. "Social Movements." In *The New Handbook of Political Sociology*, edited by T. Janoski, C. De Leon, J. Misra, and I. Martin. Cambridge: Cambridge University Press.

Jensen, K. (2002) 2020. *A Handbook of Media and Communication Research: Qualitative and Qualitative Methodologies*. 3rd ed. London: Routledge.

J.L. 2022. "Which Way, NYC-DSA?" *Cosmonaut*, November 23. https://cosmonautmag.com.

Joseph, G., and Y. Gonen. 2022. "Brooklyn Democratic Party Filed Forged Signatures to Knock Rivals Off Ballots, Fellow Dems Allege." *The City*, April 15. www.thecity.nyc.

Justice Democrats. n.d. Accessed January 16, 2023. https://justicedemocrats.com.

Kang, S., E. Oh, B. Tizol, and O. Táíwò. 2023. "You Can Win Bold Climate Laws in Your State." *Hammer & Hope* 2 (Summer). https://hammerandhope.org.

Karni, A. 2021. "A.O.C.'s Met Gala Dress Triggered Strong Reactions." *New York Times*, September 15. www.nytimes.com.

Karolidis, S., and T. Karcich. 2023. "How Did the Build Public Renewables Get Passed?" *Socialist Forum*, Spring–Summer. https://socialistforum.dsausa.org.

Karpan, A. 2022. "Progressives Fight the Machine." *Bushwick Daily*, November 7. https://bushwickdaily.com.

Kawaguchi, C. 2022. "A Tough Election for Brooklyn Democrats." *New York Daily News*, November 11. www.nydailynews.com.

Keith, D., L. March, and F. Escalona. 2023. Introduction to *The Palgrave Handbook of Radical Left Parties in Europe*, edited by D. Keith, L. March, and F. Escalona. London: Palgrave.

Kennedy, B., C. Funk, and A. Tyson. 2023. "Majorities of Americans Prioritize Renewable Energy, Back Steps to Address Climate Change." Pew Research Center, June 28. www.pewresearch.org.

Kepple, B., A. Amin, O. Egozy, J. Gross, J. Lindauer, E. Oh, and M. Kruvelis. 2022. "Resolution: Tax the Rich 2.0: Tax and Spend NYC-DSA Priority Campaign Proposal." NYC-DSA, 2022 Convention, www.nycdsacon.com.

Khlevniuk, O. 2017. *Stalin: New Biography of a Dictator*. New Haven, CT: Yale University Press.

Kitschelt, H. 2006. "Movement Parties." In *Handbook of Party Politics*, edited by R. Katz and W. Crotty. London: Sage.

Kitschelt, H. 2012. "Parties and Interest Mediation." In *The Wiley-Blackwell Companion to Political Sociology*, edited by E. Amenta, K. Nash, and A. Scott. Chichester, UK: Wiley-Blackwell.

Klein, E. 2020. *Why We're Polarized*. New York: Avid Reader.

Klein, J. 2015. "Potential Liability of Governments for Failure to Prepare for Climate Change." Sabin Center for Climate Change Law, Columbia Law School, Columbia University, New York. https://climate.law.columbia.edu.

Klein, N. 2009. "Copenhagen's Failure Belongs to Obama." *The Guardian*, December 21. www.theguardian.com.

Klein, N. 2014. *This Changes Everything: Capitalism vs. Climate*. New York: Simon and Schuster.
Klein, N. 2020. "Care and Repair: Left Politics in the Age of Climate Change." *Dissent*, Winter. www.dissentmagazine.org.
Klein, N. 2022. "How the Left Was Lost in the 1990s—but Found Its Way Again." *The Nation*, December 12. www.thenation.com.
Klinenberg, E., M. Araos, and L. Koslov. 2020. "Sociology and the Climate Crisis." *Annual Review of Sociology* 46:6.1–6.21.
Kluge, H. 2024. "Statement—Heat Claims More than 175 000 Lives Annually in the WHO European Region, with Numbers Set to Soar." World Health Organization, August 1. www.who.int.
Kolbert, E. 2014. *The Sixth Extinction: An Unnatural History*. New York: Bloomsbury.
Kose, M., N. Sugawara, and M. Terrones. 2020. *Global Recessions*. Policy Research Paper 9172. World Bank. www.worldbank.org.
Kotsonis, Y. 1999. "Review: The Ideology of Martin Malia." *Russian Review* 58 (1): 124–130.
Krase, J., and C. LaCerra. 1991. *Ethnicity and Machine Politics*. Lanham, MD: University Press of America.
Lakha, F. 2021. "Tasks for the Branches Is a Major New Project for NYC-DSA." Medium, September 24. https://medium.com.
Lakhani, N. 2022. "US Supreme Court Rejects Dakota Access Pipeline." *The Guardian*, February 22. www.theguardian.com.
Lamb, W., G. Mattioli, S. Levi, J. Roberts, S. Capstick, F. Creutzig, J. Minx, F. Müller-Hansen, T. Culhane, and J. Steinberger. 2020. "Discourses of Climate Delay." *Global Sustainability* 3 (e17): 1–5.
Lamont, M. 1994. *Money, Morals, and Manners: The Culture of the French and the American Upper-Middle Class*. Chicago: University of Chicago Press.
Lane, E., and J. Wolf. 2012. "The New York State Legislature." In *The Oxford Handbook of New York State Government and Politics*, edited by G. Benjamin. Oxford: Oxford University Press.
Lau, T. 2019. "Citizens United Explained." Brenan Center for Justice, December 12. www.brennancenter.org.
Laybourn, L., and J. Dyke. 2024. "A 'Doom Loop' of Climate Change and Geopolitical Instability." *The Conversation*, December 9. www.theconversation.com.
Leiserowitz, A., J. Carman, N. Buttermore, L. Neyens, S. Rosenthal, J. Marlon, J. Schneider, and K. Mulcahy. 2022. *International Public Opinion on Climate Change, 2022*. New Haven, CT: Yale Program on Climate Change Communication and Data for Good at Meta. https://climatecommunication.yale.edu.
Leiserowitz, A., E. Maibach, S. Rosenthal, and J. Kotcher. 2023. *Climate Change in the American Mind: Beliefs and Attitudes*. New Haven, CT: Yale Program on Climate Change Communication and George Mason University Center for Climate Change Communication. https://climatecommunication.yale.edu.
Lewis, H. 2021. "What Happened to Jordan Peterson?" *The Atlantic*, April. www.theatlantic.com.

Lewis, S. 2022. "In Defense of Campaigns." *Socialist Forum*, Summer. https://socialistforum.dsausa.org.
Limbaugh, R. 2009. "The Universe of Lies: Climate Hoax Lives, Obama to Join Copenhagen." *The Rush Limbaugh Show*, November 25. www.rushlimbaugh.com.
Lin, J. 2015. "The First Successful Climate Negligence Case: A Comment on Urgenda Foundation v. The State of the Netherlands (Ministry of Infrastructure and the Environment)." *Climate Law* 5 (1): 65–81.
Livingston, A. 2021. "Nonviolence and the Coercive Turn." In *The Cambridge Handbook of Civil Disobedience*, edited by W. Scheuerman. Cambridge: Cambridge University Press.
Louis, E. 2022. "Brooklyn's Democratic Boss Is Taking a Step Back—for the Happiest of Reasons Rodneyse Bichotte Hermelyn Isn't Done with Politics Though," *Intelligencer*, July 15. https://nymag.com.
Lynch, S. 2020. "U.S. Congresswoman Ocasio-Cortez Says Republican Colleague Called Her Profane Slur." *Reuters*, July 23. www.reuters.com.
Maisano, C. 2022. "Reflections on the NYC-DSA Convention, and Beyond." *Socialist Forum*, Fall. https://socialistforum.dsausa.org.
Malia, M. 1994. *Soviet Tragedy: A History of Socialism in Russia, 1917–1991*. New York: Free Press.
Malm, A. 2021. *How to Blow Up a Pipeline*. New York: Verso.
Malm, A., and the Zetkin Collective. 2021. *White Skin, Black Fuel: On the Danger of Fossil Fascism*. New York: Verso.
Marcetic, B. 2024. "Trump Is Planning a Third Red Scare." *Jacobin*, November 2. https://jacobin.com.
Markowicz, K. 2018. "Sorry, Democratic Socialists—You're Still Pushing Poison." *New York Post*, August 5. https://nypost.com.
Marsh, J., and N. Hicks. 2021. "Family Affair: Adams Campaign Paid Brooklyn Dem Boss' Hubby $80K." *New York Post*, November 12. https://nypost.com.
Maslow, A. 1943. "A Theory of Human Motivation." *Psychological Review* 50:370–396.
Mason, L. 2018. *Uncivil Agreement: How Politics Became Our Identity*. Chicago: University of Chicago Press.
Massoglia, A., and K. Evers-Hillstrom. 2021. "'Dark Money' Topped $1 Billion in 2020, Largely Boosting Democrats." Open Secrets, March 17. www.opensecrets.org.
Masters, B., and P. Temple-West. 2024. "Vanguard Backed No Environmental or Social Measures in 2024 Proxy Season." *Financial Times*, August 30. www.ft.com.
Mayer, J. 2016. *Dark Money: The Hidden History of the Billionaires Behind the Rise of the Radical Right*. New York: Vintage Books.
Mays, J. 2019. "A Political Party Aligns Itself with Ocasio-Cortez." *New York Times*, December 6. www.nytimes.com.
McAdam, D. (1982) 1999. *Political Process and the Development of Black Insurgency 1930–1970*. Chicago: University of Chicago Press.
McAlevey, J. 2020. *A Collective Bargain: Unions, Organizing, and the Fight for Democracy*. New York: Ecco.

McCarthy, N. 2019. *Polarization: What Everyone Needs to Know*. Oxford: Oxford University Press.

McFarlane, S. 2023. "Explainer: Global Fossil Fuel Subsidies on the Rise Despite Calls for Phase-Out." *Reuters*, November 23. www.reuters.com.

McIntosh, D. 2021. "Talking Socialism: Catching Up with AOC." *Democratic Left*, March 19. www.dsausa.org.

McNair, B. 2004. "PR Must Die: Spin, Anti-Spin and Political Public Relations in the UK, 1997–2004." *Journalism Studies* 5 (3): 325–338.

Meaker, M. 2024. "The New Face of Climate Activism Wields a Pickaxe." *Mother Jones*, August 18. www.motherjones.com.

Mellins, S. 2020. "Inside the NYC Democratic Socialists' Powerhouse Electoral Machine." *Jacobin*, October 12. https://jacobin.com.

Meyer, N. 2016. "Electoral Strategy After Bernie's Campaign." *Monthly Review* 31 (183): 10–11.

Meyer, N. 2022. "Debating Party Building at the 2022 NYC-DSA Convention." *The Call*, November 16. https://socialistcall.com.

Meyer, N. 2023. "I've been a very active member in DSA for 11 years. I've never seen the org more united than it has been in the last 6 weeks. 24 disgruntled and totally inactive members quitting a 70,000+ member organization does not a crisis make, let's please have a sense of proportion! X, November 15. https://twitter.com/nealmeyer/status/1724862877851332994.

Meyer, N. 2024. "A Big Task for the Left: Getting a Grip on the Current Moment." *Left Notes Substack*, February 1. https://leftnotes.substack.com.

Mezirow, J. 1978. "Perspective Transformation." *Adult Education* 28 (2): 100–110.

Miller, J., and A. Waggoner. 2022. "How a Fake Budget Crisis Was Used to Gut Funding for NYC Schools." NYCLU of New York, August 19. www.nyclu.org.

Milman, O. 2022. "'Rapid Acceleration' in US School Book Censorship Leads to 2,500 Bans in a Year." *The Guardian*, September 19. www.theguardian.com.

Mollenkopf, J. 2023. "The Long View: The Evolution of Flatbush." *Vital City*, April 4. www.vitalcitynyc.org.

Moody, K. 2022. *Breaking the Impasse: Electoral Politics, Mass Action, and the New Socialist Movement in the United States*. Chicago: Haymarket Books.

Moynihan, D., and N. Glazer. (1963) 1970. *Beyond the Melting Pot: The Negroes, Puerto Ricans, Jews, Italians, and Irish of New York*. Cambridge, MA: MIT Press.

Mudde, C. 2024. "The 2024 Elections: The Far Right at the Polls." *Journal of Democracy* 35 (4): 121–134.

Mufson, S. 2012. "Green Party Presidential Candidate Jill Stein Charged with Trespassing in Keystone XL Protest." *Washington Post*, October 31. www.washingtonpost.com.

Mulvey, K., S. Shulman, D. Anderson, N. Cole, J. Piepenburg, and J. Sideris. 2015. *The Climate Deception Dossiers: Internal Fossil Fuel Industry Memos Reveal Decades of Corporate Disinformation*. Union of Concerned Scientists. www.ucsusa.org.

Murphy, T. 2018. "How Alexandria Ocasio-Cortez Pulled Off the Year's Biggest Political Upset." *Mother Jones*, June 27. www.motherjones.com.

Muzzio, D. 2012. "Politics and the News Media in the Empire State." In *The Oxford Handbook of New York State Government and Politics*, edited by G. Benjamin. Oxford: Oxford University Press.

Nash, K. (2000) 2010. *Contemporary Political Sociology: Globalization, Politics, and Power*. Chichester, UK: Wiley-Blackwell.

Neilsberg Research. 2024. "Brooklyn, New York Population by Age." July 26. www.neilsberg.com.

New York Energy Democracy Alliance. n.d. "About Energy Democracy Alliance." Accessed February 6, 2025. https://energydemocracyny.org.

New York State Assembly. 2023. Bill No. S04134. "The New York State Build Public Renewables Act." February 3. https://assembly.state.ny.us.

New York State Board of Elections. n.d. "Enrollment by County." Accessed September 13, 2023. www.elections.ny.gov.

New York State Division of the Budget. 2023. "The Legislature." www.budget.ny.gov.

New York State Senate. 2000. Senate Bill 8277B. Sponsored by Jessica Ramos. May 1. www.nysenate.gov.

New York State Senate. 2023. Senate Bill S562A. "All-Electric Building Act." January 5. www.nysenate.gov.

New York State Senate. n.d. "About the New York State Senate." Accessed October 9, 2023. www.nysenate.gov.

Newport, F. 2018. "Democrats More Positive About Socialism than Capitalism." Gallup, August 13. https://news.gallup.com.

Nichols, J. 2018. "Alexandria Ocasio-Cortez Wins as a Democratic Socialist with a 21st-Century Vision." *The Nation*, June 27. www.thenation.com.

Nir, S. 2016. "Bernie Sanders Back in the Old Neighborhood to Make His Case." *New York Times*, April 8. www.nytimes.com.

Nnaemeka, S., and N. Luo. 2021. "How We Won Taxes on the Rich in New York." *The Forge*, October 21. https://forgeorganizing.org.

Nohria, N., and R. Khurana, eds. 2010. *Handbook of Leadership Theory and Practice*. Cambridge, MA: Harvard University Press.

Noor, D. 2023. "Big Oil Quietly Walks Back on Climate Pledges as Global Heat Records Tumble." *The Guardian*, July 16. www.theguardian.com.

NPC (National Political Committee of the Democratic Socialists of America). n.d. "NPC Draft—Copy for Members—2023 DSA Budget." Google Drive. https://docs.google.com.

NYC-DSA (New York City Democratic Socialists of America). 2021. "Public Power Launch Call 2021." YouTube, April 6. www.youtube.com/watch?v=ZLbJGVS7vg8&t=1173s.

NYC-DSA (New York City Democratic Socialists of America). n.d.-a. "Branches." Accessed August 9, 2024. https://socialists.nyc.

NYC-DSA (New York City Democratic Socialists of America). n.d.-b. "Our Structure." Accessed August 13, 2024. www.socialists.nyc.
NYC-DSA Steering Committee. 2023. "NYC-DSA Tasks and Perspectives of 2023." Presented to the 2023 annual NYC-DSA Conference.
NYC-DSA 2022 Convention. 2022. "Resolution: Tax the Rich and Spend." Accessed November 20, 2022. www.nycdsacon.com.
NYU Furman Center. n.d.-a. "East Flatbush." https://furmancenter.org.
NYU Furman Center. n.d.-b. "Flatbush/Midwood." Accessed September 14, 2023. https://furmancenter.org.
Ocampo, O. 2022. "An Oligarchy Expert Answers Our Questions About Wealth and Empowerment." Inequality.org, April 14. https://inequality.org.
O'Carroll, L., C. García, and P. Duncan. 2024. "Political Donations in France Swerve to the Right as Le Pen's Niece Raises More than Macron." *The Guardian*, May 30. www.theguardian.com.
Odell, J. 2019. *How to Do Nothing: Resisting the Attention Economy*. Brooklyn, NY: Melville House.
Olson, M. 1965. *The Logic of Collective Action*. Cambridge, MA: Harvard University Press.
Oxfam. 2023. *Climate Equality: A Planet for the 99%*. https://oxfamilibrary.openrepository.com.
Partanen, A. 2016. *The Nordic Theory of Everything: In Search of a Better Life*. New York: Harper.
Patel, A., and H. Imran. 2024. "The ICC Can No Longer Ignore the Genocide in Gaza." *Al Jazeera*, April 21. www.aljazeera.com.
Pettifor, A. 2019. *The Case for the Green New Deal*. London: Verso.
Piccio, D. 2019. *Party Responses to Social Movements: Challenges and Opportunities*. New York: Berghahn.
Pierre-Louis, F. 2013. "Haitian Immigrants and the Greater Caribbean Community of New York City: Challenges and Opportunities." *Memorias* 21 (Barranquilla September–December). www.scielo.org.co.
Piketty, T. 2021. "Long Live Socialism!" In *Time for Socialism: Dispatches for a World on Fire, 2016–2021*. New Haven, CT: Yale University Press.
Pistor, K. 2019. *The Code of Capital: How the Law Creates Wealth and Inequality*. Princeton, NJ: Princeton University Press.
Piven, F., and R. Cloward. 1977. *Poor People's Movements*. New York: Vintage.
Place, N. 2021. "Marjorie Taylor Greene Calls AOC a 'Scared Little Girl' as She Badgers Her to Debate Green New Deal." *The Independent*, April 23. www.the-independent.com.
Polletta, F. 2002. *Freedom Is an Endless Meeting: Democracy in American Social Movements*. Chicago: University of Chicago Press.
Polletta, F. 2020. *Ties That Bind: Imagined Relationships in Moral and Political Life*. Chicago: University of Chicago Press.
Porcelli, V., and B. Max. 2020. "Wins Pile Up for New York Left." *Gotham Gazette*, July 23. www.gothamgazette.com.

Powell, M. 2020. "A Black Marxist Scholar Wanted to Talk About Race: It Ignited a Fury." *New York Times*, August 14. www.nytimes.com.
Powell, M. 2023. "Why Older Socialists Are Quitting DSA." *The Atlantic*, November 15. www.theatlantic.com.
Public Power NY. n.d. "Member Organizations." Accessed August 22, 2024. https://publicpowerny.org.
Putnam, R. 2000. *Bowling Alone: The Collapse and Revival of American Community*. New York: Simon and Schuster.
Rabinowitz, H. (@HuntRabinowitz). 2022. Tweet. X, November 10.
Ransom, J. 2024. "Eric Adams to Stand Trial on Corruption Charges in April, Judge Says." *New York Times*, November 1. www.nytimes.com.
Raskin, S. 2022. "AOC Says She Had an 'Awakening' About Her Indigenous Heritage at Standing Rock in 2016." *New York Post*, June 4. https://nypost.com.
Ravikumar, S., and S. Twidale. 2023. "Britain Commits to Hundreds of North Sea Oil and Gas Licences." *Reuters*, July 31. www.reuters.com.
Rawls, J. (1971) 1999. *A Theory of Justice*. Rev. ed. Oxford: Oxford University Press.
Rawls, J. (1993) 2005. *Political Liberalism*. Expanded ed. New York: Columbia University Press.
Rebel Wisdom. 2018. "Addicted to Ideology? With Gabor Maté." YouTube, November 16. www.youtube.com/watch?v=x2YdpvnwtGc&t=0s.
Red Star Caucus. 2019. "Points of Unity." February 7. https://redstarcaucus.org.
Reed, A., Jr. 2000. *Class Notes: Posing as Politics and Other Thoughts on the American Scene*. New York: New Press.
Reidmiller, D., C. Avery, D. Easterling, K. Kunkel, K. Lewis, T. Maycock, and B. Stewart, eds. 2018. *USGCRP, 2018: Impacts, Risks, and Adaptation in the United States: Fourth National Climate Assessment*. Vol. 2. Washington, DC: US Global Change Research Program. https://nca2018.globalchange.gov.
Rooney, S. 2021. *Beautiful World, Where Are You?* New York: Farrar, Straus and Giroux.
Rosenthal, B. 2018. "De Blasio Donor Says He Steered Thousands in Bribes to Mayor's Campaigns." *New York Times*, March 22. www.nytimes.com.
Rothfeld, M. 2023. "Friends of Eric Adams Have Paid Fraction of Sum Ordered in Federal Case." *New York Times*, March 10. www.nytimes.com.
Saad, L. 2019. "Socialism as Popular as Capitalism Among Young Adults in U.S." Gallup, November 25. https://news.gallup.com.
Sanders, B. 2016. *Our Revolution: A Future to Believe in*. New York: Thomas Dunne Books.
Sanders, B., with H. Gutman. (1997) 2019. *Outsider in the White House*. London: Verso.
Sartori, G. 1976. *Parties and Party Systems: A Framework of Analysis*. Cambridge: Cambridge University Press.
Saunders, G. 2020. "Love Letter." *New Yorker*, March 30. www.newyorker.com.
Savage, M. 2012. "Class, Culture and Politics." In *The Wiley-Blackwell Companion to Political Sociology*, edited by E. Amenta, K. Nash, and A. Scott. Chichester, UK: Wiley-Blackwell.

Schatz, B. 2019. "A Top Republican Strategist Dismissed Alexandria Ocasio-Cortez as 'The Little Girl.' Her Response Is Perfect." *Mother Jones*, January 5. www.motherjones.com.

Schepelmann, P., M. Stock, T. Koska, R. Schüle, and O. Reutter. 2009. *A Green New Deal for Europe: Towards Green Modernisation in the Face of Crisis*. Wuppertal Institute for Climate, Environment, and Energy, Belgium. https://gef.eu.

Schlozman, D., and S. Rosenfeld. 2019. "The Hollow Parties." In *Can America Govern Itself?*, edited by F. Lee and N. McCarty. New York: Cambridge University Press.

Schneider, H., and C. Kahn. 2020. "Majority of Americans Favor Wealth Tax on Very Rich." *Reuters*, January 10. www.reuters.com.

Schrecker, E. 1998. *Many Are the Crimes: McCarthyism in America*. Boston: Little, Brown.

Sebestyen, V. 2018. *Lenin: The Man, the Dictator, and the Master of Terror*. New York: Vintage.

Sen, A. 1999. *Development as Freedom*. Oxford: Oxford University Press.

Setzer, J., C. Higham, and E. Bradeen. 2022. "Challenging Government Responses to Climate Change Through Framework Litigation." London School of Economics, September 7. www.lse.ac.uk.

Siegel, J., and J. Bram. 2024. "Spotlight: New York City's Rental Housing Market." Office of New York City Comptroller Brad Lander, January 17. https://comptroller.nyc.gov.

Sinnott, S., and P. Gibbs. 2014. *Organizing: People, Power, Change*. The Commons: Social Change Library. https://commonslibrary.org.

Skeggs, B. 1997. *Formations of Class and Gender*. London: Sage.

Skocpol, T. 2003. *Diminished Democracy: From Membership to Management in American Civic Life*. Norman: University of Oklahoma Press.

Skocpol, T., and C. Tervo. 2020. *Upending American Politics: Polarizing Parties, Ideological Elites, and Citizen Activists from the Tea Party to the Anti-Trump Resistance*. New York: Oxford University Press.

Small, M., and J. Calarco. 2022. *Qualitative Literacy: A Guide to Evaluating Ethnographic and Interview Research*. Berkeley: University of California Press.

Smith, N. 2021. "Interview: Saikat Chakrabarti, Creator of the Green New Deal." *Noahpinion Substack*, February 28. https://noahpinion.substack.com.

Snow, D., E. Rochford Jr., S. Worden, and R. Benford. 1986. "Frame Alignment Processes, Micromobilization, and Movement Participation." *American Sociological Review* 51:464–481.

Snyder, T. 2010. *Bloodlands: Europe Between Hitler and Stalin*. New York: Basic Books.

Solnit, R. 2019. "Standing Rock Inspired Ocasio-Cortez to Run. That's the Power of Protest." *The Guardian*, January 14. www.theguardian.com.

Stein/Baraka. 2016. "The Green New Deal Is a Job Creator." Jill Stein 2016 presidential campaign video. YouTube, November 3. www.youtube.com/watch?v=PGDjnUXmWFM&t=87s.

Steinbeck, J. 1939. *The Grapes of Wrath*. New York: Penguin.

Stephens, B. 2018. "Democratic Socialism Is Dem Doom." *New York Times*, July 6. www.nytimes.com.

Stetler, H. 2024. "Corporate France Is Making Peace with Marine Le Pen." *Jacobin*, June 27. https://jacobin.com.
Stockemer, D., and A. Sundström. 2023. "Age Inequalities in Political Representation: A Review Article." *Government and Opposition: An International Journal of Comparative Politics*, May 3, 1–18.
Stockholm International Peace Research Institute. 2024. "Role of Nuclear Weapons Grows as Geopolitical Relations Deteriorate—New SIPRI Yearbook Out Now." June 17. www.sipri.org.
Stout, J. 2010. *Blessed Are the Organized: Grassroots Democracy in America*. Princeton, NJ: Princeton University Press.
Talmadge, E. 2023. "Citizens' Assemblies: Are They The Future of Democracy?" *The Guardian*, February 1. www.theguardian.com.
Tarleton, J. 2015. "All Systems Go at Mayday Community Space." *The Indypendent*, November 11. https://indypendent.org.
Tarrow, S. 2021. *Movements and Parties: Critical Connections in American Political Developments*. Cambridge: Cambridge University Press.
Taylor, K.-Y. 2020. "Reality Has Endorsed Bernie Sanders." *New Yorker*, March 30. www.newyorker.com.
Temple-West, P. 2023. "Shareholders Raise Heat on US Companies over Executive Pay." *Financial Times*, December 14. www.ft.com.
Temple-West, P. 2024. "Proxy Season Results Show Support for ESG Efforts Continues to Ebb." *Financial Times*, July 5. www.ft.com.
Thier, H. 2023. "I'm a Proud Jewish DSA Member: Here's Why I'm Not Quitting." *The Nation*, October 25. www.thenation.com.
Thunberg, G. 2019. "'Our House Is on Fire': Greta Thunberg, 16, Urges Leaders to Act on Climate." *The Guardian*, January 25. www.theguardian.com.
Tilly, C. 1978. *From Mobilization to Revolution*. New York: McGraw-Hill.
Tilly, C. 1986. *The Contentious French*. Cambridge, MA: Harvard University Press.
Tolentino, J. 2020. "Barbara Ehrenreich Is Not an Optimist, but She Has Hope for the Future." *New Yorker*, March 21. www.newyorker.com.
Tufekci, Z. 2018. *Twitter and Tear Gas: The Power and Fragility of Networked Protest: The Power and Fragility of Networked Protest*. New Haven, CT: Yale University Press.
United Nations, Department of Economic and Social Affairs. 2009. "A Global Green New Deal." https://sdgs.un.org.
Urban Institute. 2024. "Nine Charts About Wealth Inequality in America." April 25. https://apps.urban.org.
US House of Representatives. 2019. H. Res. 109 (The Green New Deal Resolution). Proposed by Alexandria Ocasio-Cortez. February 7. www.congress.gov.
US House of Representatives. 2022. H.R. 5376. Public Law No. 117-169. August 16. www.congress.gov.
US Senate. 2022. Inflation Reduction Act of 2022. Proposal for H.R. 5376. www.democrats.senate.gov.

US Senate. n.d. "Types of Legislation." Accessed February 2, 2023. www.senate.gov.
Uteuova, A. 2023. "New York Takes a Big Step Toward Renewable Energy in 'Historic' Climate Win." *The Guardian*, May 3.
Vachon, T., M. Wallace, and A. Hyde. 2016. "Union Decline in a Neoliberal Age: Globalization, Financialization, European Integration, and Union Density in 18 Affluent Democracies." *Socius: Sociological Research for a Dynamic World* 2:1–22.
Van Auken, T. 2020. "Organizing in the Age of Coronavirus: How to Campaign in a Crisis." *The Forge*, June 12. https://forgeorganizing.org.
Walder, A. 2009. "Political Sociology and Social Movements." *Annual Review of Sociology* 35:393–412.
Walker, S. 2022. "The Meaning and Potential of a Human Rights-Based Approach to Climate Change Post-Sharma." *Alternative Law Journal* 47 (3): 194–198.
Warren, E., and N. Williams. 2022. "Senator Warren, Rep. Williams to Introduce Bill to Expand Youth Access to Voting." Press release, July 11. www.warren.senate.gov.
Weber, M. (1921) 2009. "Politics as Vocation." In *From Max Weber: Essays in Sociology*, edited by H. Gerth and C. Mills. London: Routledge.
Weiss, B. 2018. "Julia Salazar, the Left's Post-Truth Politician." *New York Times*, September 14. www.nytimes.com.
Wilson, J. 1962. *The Amateur Democratic: Club Politics in Three Cities*. Chicago: University of Chicago Press.
Wilson, S. 2023. "Gentrification by Fire." *Washington Post*, February 10. www.washingtonpost.com.
Winters, J. 2011. *Oligarchy*. Cambridge: Cambridge University Press.
World Economic Forum. 2018. "These Are the Cities with the Biggest Carbon Footprints." July 2. www.weforum.org.
World Meteorological Organization. 2022a. *State of the Climate in Africa 2021*. Geneva: World Meteorological Organization. https://library.wmo.int.
World Meteorological Organization. 2022b. *State of the Climate in Latin America and the Caribbean 2021*. Geneva: World Meteorological Organization. https://library.wmo.int.
Wright, E. 2019. *How to Be an Anticapitalist in the Twenty-First Century*. London: Verso.
Yeampierre, E. 2020. "Making the Green New Deal the Real Deal." In *AOC: The Fearless Rise and Powerful Resonance of Alexandria Ocasio-Cortez*, edited by L. Lopez. New York: St. Martin's.
Zimmerman, A. 2022. "Eric Adams Is Facing Pressure to Reverse NYC School Budget Cuts. Should He?" *Chalkbeat New York*, August 4. https://ny.chalkbeat.org.

INDEX

Page numbers in italics indicate Figures, Tables, or Photos

Ackerman, Seth, 186–87
Adams, Eric, 95, 101. *See also* machine politics
affective and moral commitments. *See* moral commitments and engagement, of NYC-DSA
Alexis, David, 22, 77, 105–6, 121; recruitment as candidate, 107–8
Alexis, David, campaign for New York State Senate (2022), 24–25; canvassing experience in, 190; communications about "Public Power" and BPRA from, 118, 222–23; Democratic Party and Brooklyn elites relationship with, 91–111; ESWG relationship with, 124–25, 137–38, 169, 218, 220; NYC-DSA leaders in, 169–77; organization, 112–42; "Public Power" electoralization by, 27, 218–22, 224, 226, 227, 230; social media communications of, 118; stakes, 112, 220. *See also* canvassers, in Alexis campaign; volunteers, in Alexis campaign
Alinsky, Saul, 19, 155
ambition and seriousness, in political organizing, 161, 163–65, 175
anticlimate movement, 39–40
anticommunism, 5, 9, 15–18, 26, 43, 81, 85–88
anti-Trump Resistance movement, 13, 79; scholarship, 52–53
AOC. *See* Ocasio-Cortez, Alexandria

apathy, 1, 4, 121. *See also* political participation
apolitical understandings of climate change, 39
author biography and ethnographic experience, 16–17, 115. *See also* ethnography

barriers, to political participation, 41–46
Beautiful World, Where Are You? (Rooney), 88–89
"Bernie Bros," in NYC-DSA, 72
Bernstein, Eduard, 16, 177
Bichotte Hermelyn, Rodneyse, 95–96, 101, 105. *See also* machine politics
Biden, Joe, 133
Bigger Than Bernie (Day and Uetricht), 21, 96–97
Blair, Tony, 86
"A Blueprint for a New Party" (Ackerman), 186–87
Bourdieu, Pierre, 45, 50
Bowman, Jamaal, 65, 207, 232–33
BPRA. *See* Build Public Renewables Act
Brand New Congress, 78, 199
Brooklyn, 75; Alexis campaign in, 91–111; as center for the Left and NYC-DSA, 9, 10, 11, 21, 132, 179; Central Brooklyn Electoral Working Group, 11–12, 149, 182, 184–85; demographics, 98, 102–4, 122; DSA-backed electeds in, 65; DSA branches in, 24, 149, 179; music field in, 126; Park Slope neighborhood of,

Brooklyn (*cont.*)
   22, 91, 103, 109, 115, 212; political field, 95, 98–111; after Trump victory in 2016, 77, 109–10. *See also* Democratic Party in Brooklyn; Flatbush, Brooklyn
Brooklyn Socialist Club, 11
Brown, Wendy, 50–51
Brulle, Robert, 39
Build Public Renewables Act (New York State) (BPRA), 195, 226, 227, 233; Alexis campaign communication about, 118, 222–23; electoralization of, 219–20, 230; ESWG and Public Power NY Coalition and, 210–15, *213*; media coverage of "Public Power" and, 193, 216, 223; Parker delaying, 112; provisions of, 192–93; support for, 218–19
Bush, Cori, 232

campaign organizing, 24, 151–56; candidates role in, 56, 180; Van Auken on, 66–67, 74–75, 159–60. *See also* Alexis, David, campaign for New York State Senate; leadership, in NYC-DSA; leadership theory
candidates: campaign organizations role of, 56, 180; movement, 182–83; primary challenges, 14, 71, 108, 144, 206, 220; recruitment in NYC-DSA, 160–62; social capital of NYC-DSA, 107–8, 126; transition from movement sphere to institutional sphere, 165–68; young, 44
canvassers, in Alexis campaign, 122–24; casual, 130–34; organizer, 137–40; pundit, 134–36
canvassing, 116–17, 121–22, 177; climate and "Public Power," 221, 223–24; as core collective practice in NYC-DSA, 113, 118, 185; media communication contrasted with, 135–36, 191, 238–40; opening ritual, 118–19; photo ritual, 120; political citizenship strengthened with, 120, 135; social media communication supporting, 118
capitalism, 10, 16, 47, 52, 89, 130–31, 240; climate change and, 39, 80, 129; democratic socialism contrasted with, 17, 84–88; Great Recession and, 5, 51. *See also* neoliberal capitalism
Central Brooklyn Electoral Working Group, 11–12, 149, 182, 184–85
Chakrabarti, Saikat, 199
Chomsky, Noam, 46
Citizens United (Supreme Court decision) (2010), 241
Citywide Leadership Committee, 149–50, 170
Citywide Steering Committee, 146, 149–50, 170
civil rights movement, 176–77
class. *See* race, ethnicity, and class
CLCPA. *See* Climate Leadership and Community Protection Act
cleavages, political and social, 61
climate change: apolitical understandings of, 39; capitalism and, 39, 80, 129; media interest in, 221
Climate + Community Project, 206, 215
climate crisis, 1, 232; democracy and, 2–4, 44; experience of, 23, 44, 116, 221, 223; Green New Deal resolution and, 200; millennial urban experience of, 88–90; movement capacity built in, 227; politicization of, 40, 221; scholarship for survival, 242–44; state role in, 3–4, 32. *See also* Ecosocialist Working Group
climate justice, 192, 193, 201, 210, 212, 234, 235
Climate Leadership and Community Protection Act (CLCPA) (2019), 214, 215
climate legislation, 14, 93, 95, 192, 194, 214, 215, 226. *See also* Build Public Renewables Act; Green New Deal
climate movement, 13, 26; Green New Deal movement and, 29, 195, 210;

movement-centrism, 35–36; NYC-DSA and, 112, 124, 125, 127, 170, 194–95, 202–27, 213, 217, 226; politics relationship to, 31–34, 39–41, 235; protest-based organizing, 36–38; radical flank, 34
climate neglect, by government, 32
climate organizing, 127–28, 207, 208–10, 223–24
climate sociology, 39
Clinton, Bill, 71, 86
Clinton, Hillary, 71, 73, 79, 133, 176, 183
clubs. *See* Democratic clubs
coercive turn, 36
Cold War. *See* anticommunism
commitments, moral. *See* moral commitments and engagement, of NYC-DSA
communism, 2, 3, 16; dictatorships and, 52, 84–85. *See also* anticommunism
Conference of the Parties (COP), 32
consultants, political, 180–81, 182
corruption, 101–2
COVID-19 pandemic, 91, 155; economic inequality and, 16; NYC-DSA impacted by, 13, 135, 156, 164; research impacted by, 22, 114, 115
crisis of democracy. *See* democracy, crisis of
cultural turn, 50, 59

Danish Agriculture and Food Council, 37, 38
dark money, 241
Day, Meagan, 21, 72, 96–97
Della Porta, Donatella, 53, 56, 57–58, 61–62
democracy. *See* Democratic Socialists of America; grassroots democracy; leadership, in NYC-DSA; neoliberal capitalism; New York City Democratic Socialists of America
democracy, crisis of, 4–5, 57; Democratic clubs and, 95–96, 98–100, 110, 117, 183–84; in US, 8, 13, 228, 232–33. *See also* Democratic Party; Democratic Party in Brooklyn; movement parties; political participation
Democratic clubs, 95–96, 98–100, 110, 117, 183–84
Democratic Party: democratic decline of, 152, 159, 179, 232; democratic socialism as alternative to centrist establishment in, 79, 82, 176; DSA as moral alternative to, 27, 80, 83, 96, 101, 102; the Left fear of cooption by, 145, 148, 163, 187; moral failures, 67, 96, 101–2, 215–17, 217; in New York State, 99–102, 103; NYC-DSA as independent faction of, 186, 188, 230; NYC-DSA fears of cooption by, 97, 187; NYC-DSA feelings on, 42, 79; political organization, 152, 180–81; Sanders insurgency in, 71, 96, 175. *See also* leadership, in Democratic Party
Democratic Party in Brooklyn: democratic decline of, 13–14, 95; leadership, 95–96, 103; NYC-DSA as departure from, 9, 13, 27, 78; NYC-DSA perception from, 95; NYC-DSA using ballot in, 14, 167; organizational history of, 98–102, 103, 108, 183; progressive faction in, 108–11; in Trump era, 109–10
democratic socialism, 26–27, 34; capitalism contrasted with, 17, 84–88; civil rights movement and, 176–77; communism contrasted with, 2, 3, 16; concept and label, 16–18, 22, 80–81, 82, 83, 84, 127–28, 177; elections as tactic for, 112, 184–85, 206, 216, 220, 222; Great Recession movements for, 2, 5–8, 10–12, 20, 53–55, 58, 82; green parties contrasted with, 6, 197, 199–202; millennial urban experience, 88–90; movement parties, 2, 5–8, 26; patriarchy, 145; revival of, 15–17, 20, 81, 84–86; stigma of, 5, 15–18, 87–88, 146; universalism in, 6, 55, 83, 225, 232; welfare and, 16, 94

Democratic Socialist Organizing Committee, 144–45
Democratic Socialists of America (DSA): challenges to cohesion and party identity, 158, 188–90; electeds in New York State government, 65, 166–67; history, 144–48; media communication inside, 157–58; membership, 13; as moral alternative to Democratic Party, 27, 80, 83, 96, 101, 102; National Political Committee, 146, 157–58; New York City as center for, 9, 10, 11, 13, 21, 98, 117, 132, 172, 179; patriarchy, 145; resurgence of, 9–11, 13, 63–64; Sanders impact in, 9–13, 71–76, 82; Trump victory in 2016 impact on, 76–79. *See also* green parties; New York City Democratic Socialists of America; Occupy Wall Street; Working Families Party
democratization, of political organization, 178–82. *See also* leadership, in NYC-DSA
dirty break strategy, 97, 186, 189
disembedment. *See* reterritorialization of political life
DSA. *See* Democratic Socialists of America; New York City Democratic Socialists of America
Duhalde, David, 78, 190

economic inequality, 4, 5, 16, 51–53, 61, 71
Ecosocialist Working Group (ESWG), 12, 25, 150; Alexis campaign and, 124–25, 137–38, 169, 218, 220; BPRA and Public Power NY Coalition and, 210–15, *213*; environment for climate politics in New York and, 202–10; goals, 193, 211, 214, 223; Gordillo and history of, 204, 205–6; organization and organizing in Flatbush, 24, 101; "Public Power" campaign escalation and, 216–26, *217*, *226*; successes and failures, 107–8, 193, 194–95, 215–16, 227, 228, 229–31, 234

Ehrenreich, Barbara, 12, 145, 147–48
elections, as democratic socialist movement tactic, 112, 184–85, 206, 216, 220, 222. *See also* mobilizing structures
electoralization: BPRA, 219–20, 230; "Public Power," 27, 218–22, 224, 226, 227, 230
electoral work and electoral campaigns, 12, 24–25. *See also specific candidates and elections*
elite domination, in political field, 4, 7, 8, 9, 13, 46, 98, 216, 232. *See also* democracy, crisis of; neoliberal capitalism
Energy Democracy Alliance, 206, 214
entryism, 157
ESWG. *See* Ecosocialist Working Group
ethnicity. *See* race, ethnicity, and class
Ethnicity and Machine Politics (Krase and LaCerra), 98
ethnography, 20–21, 24–25; cognitive empathy, 114; transformative ethnographic experience, 23. *See also* field research
European welfare states, 16, 55
exclusion, from political field, 45
Extinction Rebellion, 32–33, 34, 36–38, 40–41

failures. *See* successes and failures
Feaver, Julianne, 223, 225
feminism, 145
field research, 20; author journey in, 22–25; embodied and digital, 21
Fisher, Dana, 35, 52–53, 152
Five Star Movement (Movimento 5 Stelle), 54
Flatbush, Brooklyn, 91–92, 116, 126; climate consciousness, 127, 223–24; Haitian diaspora in, 77, 103, 104, 105; NYC-DSA organization, 24, 101, 110, 113, 115, 121, 124, 138. *See also* Alexis, David, campaign for New York State Senate
Fox News, 88

Fridays for Future NYC, 40, 44
Friedman, Thomas, 196
Fukuyama, Francis, 84–85

Galbraith, John Kenneth, 80
Galvis, Tefa, 169, 171–74
Ganz, Marshall, 143, 151–56, 163, 243
gender, 18, 108; balance and diversity in NYC-DSA, 123, 161, 203; "Bernie Bros," in NYC-DSA and, 72; gendered bullying and, 88; patriarchy in democratic socialism, 145
genocide, in Palestine, 13, 190, 233, 234
gerontocracy, 43–45
Gibbs, Peter, 154
Giddens, Anthony, 19, 86, 100
goals: ESWG, 193, 211, 214, 223; long- and short-term capacity and, 70, 152, 156, 193, 204, 211, 228, 230, 231, 242; NYC-DSA, 144–49, 159, 160, 161, 163–65, 177, 185, 193
Gordillo, Gustavo, 169, 170–71; ESWG history and, 204, 205–6; political organization, 167, 182
Goulden, Daniel, 124–25, 137–38
Graeber, David, 51–52
The Grapes of Wrath (Steinbeck), 47
grassroots democracy: evolution of, 151–52; Ganz and, 152–53; history, 6–7, 19; NYC-DSA and, 14, 19, 149–50, 159, 165–68, 178–84; Sanders campaign lacking, 74. *See also* machine politics
Great Recession: democratic socialist movements of, 2, 5–8, 10–12, 20, 53–55, 58, 82; right-wing movements, 56; skepticism about capitalism and inequality and, 5, 51; transformation of political field and, 53, 56, 61, 65, 66, 242
green. *See* visual identity
Green New Deal, 193, 212; AOC, 197–99; climate movement and, 29, 195, 210; congressional resolution, 28–29, 199–202; Green Party, 196–97

Green New York design system, 224–25
green parties, 7; democratic socialism contrasted with, 6, 197, 199–202; Green Party in US, 186, 196–97
Gross, Jack, 163–65
Guinn, Andrea, 84, 158, 174, 220
Gunn-Wright, Rhiana, 199

Habermas, Jürgen, 31
Haitian diaspora, in Flatbush, 77, 103, 104, 105
Harrington, Michael, 12, 145, 147
Harris, Kamala, 15, 79, 231
Harvey, David, 50–51
Hawkins, Howie, 196–97
hierarchy of needs, 23
How Social Movements Can Save Democracy (Della Porta), 57
How to Do Nothing (Odell), 89–90
Hurricane Sandy, 221

ideology, 54, 56, 81, 95, 234–35; end of, 85; NYC-DSA, 17–18, 162–65, 174–77, 182; racist, of Trump, 77. *See also* democratic socialism; political organization/organizing
The Indypendent (newspaper), 73
inequality, economic, 4, 5, 16, 51–53, 61, 71
Inflation Reduction Act (2022), 226, 195, 207, 232
institutional sphere. *See* movement sphere to institutional sphere
IPCC 2018 (UN report), 200–201

Jacobin (magazine), 21, 68, 83, 115; reading groups, 11, 69, 73, 147
Jacobs, Rhoda, 101, 108–9
Jasper, James, 42, 60
Justice Democrats, 199

King, Martin Luther, Jr., 83, 177
Kitschelt, Herbert, 55–56, 57–58
Klein, Naomi, 5, 32, 89, 204–5, 206

Koestler, Arthur, 85
Krase, Kathryn, 98, 109–10
Kraushaar, Josh, 71, 117, 119–20, 139–40

LaCerra, Charles, 98
Lakha, Aina, 79, 121, 161, 170, 174–77
@LarryWebsite, 84
leadership, in Democratic Party: Brooklyn, 95–96, 103; national, 71; New York State, 101–2
leadership, in NYC-DSA, 143; in Alexis campaign, 169–77; campaign, 159–60; exercise of power, 176; Ganz compared with, 163; internal recruitment of, 160–62; macrolevel challenges, 156, 188; movement and electeds coordination by, 165–68; "old" and "new," 144–48; personal qualities of, 162, 183; philosophy and theory of, 151–56, 162–65, 176–77; selection of, 160–62; volunteers elevated into, 66, 159–60
leadership theory: distributed, 154; macrolevel challenge, 156, 188; NYC-DSA, 151–56, 162–65, 176–77; snowflake, 154. *See also* political organization/organizing
"Leading Change" (Ganz), 153–55
the Left. *See specific topics*
"left of the possible," 145
legislation, 166–68; climate, 14, 93, 95, 192, 194, 214, 215, 226; Inflation Reduction Act, 226, 195, 207, 232. *See also* Build Public Renewables Act; Green New Deal
Lenin, Vladimir, 3, 175, 176
Letzte Generation, 34
Lewis, Sam 12–13, 164, 169, 181–82
Litman, Amanda, 44
The Logic of Collective Action (Olson), 59
losses. *See* successes and failures

machine politics, 9, 98, 99–102, 108, 110
macrolevel challenges, organizational, 156, 188
Madison Club, 98, 99

Malm, Andreas, 35
Mamdani, Zohran, 12, 65, 218
Manchini, Helen, 44
Marx, Karl, 3, 10, 16, 81
Marxist revival, 51–53
Maslow, Abraham, 23
mass media. *See* media
Mayday Space, 11
McAlevey, Jane, 15, 97–98, 243
McCooey, John, 99, 100, 103
McManus, Devin, 78, 116, 121, 124–25, 126–28
media, 63, 88, 100; canvassing contrasted with, 135–36, 191, 238–40; climate change interest of, 221; communication inside DSA, 157–58; environment of politics, 238–40; NYC-DSA members diet of, 127, 129, 130, 132, 133–34, 136, 140, 171; ownership, 46, 171; political organization and, 152, 155–56, 191, 204, 223, 233; "Public Power" and BPRA coverage by, 193, 216, 223; socialist movement, 83–84; theory, 155–56. *See also* social media
methodology, 20–21
Meyer, Neal, 68–69, 73, 146, 186
Mezirow, Jack, 15–16
mobilizing structures, 58–60, 149, 173, 185, 231
Moody, Kim, 15, 21–22
moral alternative, DSA as, 27, 80, 83, 96, 101, 102
moral commitments and engagement, of NYC-DSA, 42, 122, 177, 183, 230
moral failures: climate neglect in government and, 32; Democratic Party, 67, 96, 101–2, 215–17, 217
movementization, 7–8
movement parties: democratic socialist, 2, 5–8, 26; the far Right and, 7–8, 43, 239; history and theory, 6–8, 19, 26, 55–58; NYC-DSA, 14, 148–50, 156, 157, 168, 178, 186, 194, 218, 228, 230

movement sphere to institutional sphere: transition of, 165–68; translation of, 215
movement theory. *See* social movement theory
Movimento 5 Stelle (Five Star Movement), 54
music field, in Brooklyn, 126

National Political Committee, DSA, 146, 157–58
neoliberal capitalism, 50–51, 66; parties and political organization in, 8, 55–57, 61, 64, 230, 240–42; revival of socialism and, 84–86; welfare decline in, 4, 5, 54, 86
New American Movement, 144–45
New Deal, 196. *See also* Green New Deal
A New Era of Public Power (Climate + Community Project 2021 report), 215
news media. *See* media
New York City: as center for Democratic Party, the Left, and DSA, 9, 10, 11, 13, 21, 98, 117, 132, 172, 179; environmental social science and, 80; neighborhood demographics, 102–3; social history, 8–9. *See also* Brooklyn; Flatbush
New York City Democratic Socialists of America (NYC-DSA): AOC impact on, 12, 82, 97, 126, 133, 138, 139–40, 172; AOC on candidates of, 182–83; ballot in Democratic Party in Brooklyn, 14, 167; canvassing, 113, 118, 185; caucuses, 150, 177, 189; challenges to cohesion and party identity, 158, 188–90; climate organizing, 112, 124, 125, 127, 170, 194–95, 202–27, 213, 217, 226; COVID-19 pandemic impact on, 13, 135, 156, 164; Democratic Party contempt and dissatisfaction from, 42, 79; Democratic Party cooption fears in, 97, 187; as Democratic Party in Brooklyn alternative, 9, 13, 27, 78; Democratic Party in Brooklyn perceptions of, 95; as Democratic Party independent faction, 186, 188, 230; democratization of political campaigns, 178–82; demographic, 72, 122, 123, 146, 161, 203; dirty break strategy, 97, 186, 189; in Flatbush, 24, 101, 110, 113, 115, 121, 124, 138; gender and, 72, 123, 161, 203; goals, 144–49, 159, 160, 161, 163–65, 177, 185, 193; history and transformation, 5, 10–15, 144–50, 159, 189; idealism, 107, 117; ideology and organization, 17–18, 162–65, 174–77, 182; international context, 13; long- compared with short-term organizational capacity and goals, 70, 152, 156, 193, 204, 211, 228, 230, 231, 242; media diet of members, 127, 129, 130, 132, 133–34, 136, 140, 171; membership, 13; moral commitments, 42, 122, 177, 183, 230; movement culture and movement dimension in, 2, 14, 58, 82, 94, 95, 115, 118, 163, 193, 224–25, 231, 241; movement party, 14, 148–50, 156, 157, 168, 178, 186, 194, 218, 228, 230; on Obama, 66–67, 74–75, 133, 159–60; Obama 2008 campaign impact on, 65, 66, 159, 178; offices, 11; "old" and "new," 9, 10–15, 24, 28, 63, 98, 100, 110, 144–50, 159, 189; party dimension and identity, 158, 184, 186–91; party system position of, 186–91; political organization/organizing in, 66, 143–48, 159–60, 163, 165–68, 176–77, 183; professionals participation in, 206, 225; Sanders influence and impact on, 10–12, 65, 72–76, 79, 96, 128, 133, 135, 149, 150, 175, 176, 184, 188; social capital of candidates, 107–8, 126; successes and failures, 15–16, 107–8, 229–31, 234; Trump 2016 victory response from, 77–79, 126, 133, 136, 138, 176, 198; visual identity, 84, 115, 118, 122–23, 141, 146. *See also* leadership, in NYC-DSA

New York Power Authority (NYPA), 193, 210–11, 212, 215
New York State, 9, 12, 113, 229; Democratic Party in, 99–102, 103; DSA electeds in, 65, 166–67; energy transition in, 192–93, 194–95, 206, 209, 211, 220; Green New York design system and, 224–25. *See also* Alexis, David, campaign for New York State Senate; Brooklyn; Build Public Renewables Act; Flatbush; New York City
New York Times, 76, 88, 134, 140, 207
Niemeyer, Marsha, 70
#NoDAPL movement, 198
The Nordic Theory of Everything (Partanen), 23–24
NYC-DSA. *See* New York City Democratic Socialists of America
NYPA. *See* New York Power Authority

Obama, Barack: climate politics, 32; NYC-DSA members on, 66–67, 74–75, 133, 159–60; Sanders campaign compared with, 74–75; 2008 campaign and victory, 19, 74, 152–53; 2008 campaign impact on NYC-DSA, 65, 66, 159, 178
Ocasio-Cortez, Alexandria (AOC), 65; climate politics, 197–99; critique of Democratic Party in New York State, 101–2; on NYC-DSA candidates, 182–83; political vision and definition of socialism, 82–83, 232; symbolic and physical violence against, 88; 2018 electoral victory impact on NYC-DSA, 12, 82, 97, 126, 133, 138, 139–40, 172
Occupy Wall Street, 9; history, 67–71; impact, 10, 11, 65, 67–71, 74, 138
Odell, Jenny, 89–90
oligarchy, 187, 241, 242
Olson, Mancur, 59
"1-2-3-4 Plan," 188–89
opportunity structures, 35, 59, 65, 195, 214. *See also* social movement theory

organization. *See* leadership, in NYC-DSA; leadership theory; political organization/organizing
organizational capacity and goals. *See* goals; political organization/organizing
Organizing (Ganz), 153–55

Palestine, 13, 190, 233, 234
pandemic. *See* COVID-19 pandemic
Parker, Kevin, 105, 107; campaign communications, 118; deception by, 215; NYC-DSA pressuring and primarying, 27, 101, 112, 220
Park Slope (Brooklyn neighborhood), 22, 91, 103, 109, 115, 212
Partanen, Any, 23–24
party identity, 61; NYC-DSA, 158, 184, 186–91
party organizations, 55–57, 179–80; movementization of, 7–8
patriarchy, 145
Paulson, Michael, 68, 214, 220–22
Piketty, Tom, 52
pipeline fights, 129, 198, 207, 209, 210
Podemos, 55
polarization, 42–43
political cleavages, 61
political consultants, 180–81, 182
political education, 106–7, 134, 161; Ecosocialist Working Group, 204
political organization/organizing, 6, 238; ambition and seriousness in, 161, 163–65, 175; campaigns and, 24, 151–56; in climate crisis, 2, 221, 223–24; democratization of, 178–82; long- and short-term capacity and goals compared, 7, 15, 31, 33–34, 70, 152, 156, 193, 204, 211, 228, 230, 231, 242; neoliberal capitalism and, 8, 55–57, 61, 64, 230, 240–42; in NYC-DSA, 66, 143–48, 159–60, 163, 165–68, 176–77, 183; reactive compared with proactive, 70, 71, 172,

195; Sanders, 74–75; theory of, 18–19. *See also* Ecosocialist Working Group; leadership; movement parties; party organizations
political participation: declining, 31, 41–46; visual identity stimulating, 122–23, 141
political sociology, 18, 26, 48–53, 237–38, 242–44
Pollak, Michael, 135–36, 223–24
primary elections. *See* elections, as democratic socialist movement tactic; *specific candidates and elections*
"Proposal for Brooklyn Branch DSA Electoral Committee" (2016 NYC-DSA proposal), 184–85
"Public Power" campaign, 24, 112, 124, 125, 194, 195, 197; Alexis campaign and electoralization of, 27, 218–22, 224, 226, 227, 230; Alexis campaign communication about, 118, 222–23; development of, 210–15, *213*; escalation of, 216–26, *217*, *226*; media coverage of, 193, 216, 223
Public Power NY Coalition, 195, 214, 215, 218

race, ethnicity, and class, 102–9, 171, 174, 209, 212; civil rights movement and, 176–77; in NYC-DSA, 72, 122, 123, 146, 161, 203; vulnerable populations and, 193, 200–201, 202
racial justice movement, 83, 173, 174, 177, 201
Racimo, Fernando, 36, 37–38
racism, 77, 171
Rawls, John, 36
reactive compared with proactive organizing, 70, 71, 172, 195
reading groups, *Jacobin*, 11, 69, 73, 147
realignment, 145, 148, 186
recruitment: of Alexis as candidate, 107–8; of NYC-DSA candidates and leaders, 160–62

reembedment. *See* reterritorialization of political life
Reibstein, Sarah, 121, 130–32
Resistance. *See* anti-Trump Resistance movement
resource mobilization theory (RMT), 58–60, 62, 63. *See also* political sociology
reterritorialization of political life, 19, 100, 204, 239
revivals: of Marxism, 51–53; of socialism, 15–17, 20, 81, 84–86
revolutionary socialism, 16
Right, the far, 7–8, 43, 239. *See also* Trump, Donald
RMT. *See* political sociology; resource mobilization theory
Rooney, Sally, 88–89
Rudebusch, William, 132–34
Run for Something, 44
Russian Revolution, 16

Salazar, Julia, *65*; transition from movement sphere to institutional sphere, 166–68; 2018 electoral victory impact, 12, 97, 138, 175
Sanders, Bernie: democratic socialism revival and, 16; impact on DSA, 9–13, 71–76, 82; influence and impact on NYC-DSA, 10–12, 65, 72–76, 79, 96, 128, 133, 135, 149, 150, 175, 176, 184, 188; Obama campaign compared with, 74–75; political organization, 74–75; political project 10, 71–72, 81–82, 96; presidential primaries, 10, 71–72, 96, 148
Schröder, Gerhard, 86
Scientist Rebellion, 36–38. *See also* Extinction Rebellion
"second floor," 168
senate campaign. *See* Alexis, David, campaign for New York State Senate
seriousness and rigor, in political organizing, 161, 163–65, 175

Shrestha, Sarahana, 65, 218, 219, 220
Sierra Club, 209
Sinnott, Shea, 154
Snow, David, 59
social cleavages, 61
socialism. *See* democratic socialism
Socialists in Office Committee, 166–67
social media, 7, 155–56, 158, 239; canvassing supported by, 118; Twitter/X, 84, 127, 134, 136, 139, 140
social movement theory, 35–36, 58–60, 63; democratic socialism and, 2, 16. *See also* leadership theory; political organization/organizing
sociology: climate, 39; climate survival project, 237, 243–44; Marxist revival, 51–53; of modernity and globalization, 100; political, 18, 26, 48–53, 237–38, 242–44
Standing Rock, 198. *See also* pipeline fights
Stein, Jill, 197
Steinbeck, John, 47
Steinem, Gloria, 145
stigma, of socialism, 5, 15–18, 87–88, 146. *See also* anticommunism
Stout, Jeffrey, 19
successes and failures: ESWG, 107–8, 193, 194–95, 215–16, 227, 228, 229–31, 234; ESWG goals relation to, 193, 211, 214, 223; NYC-DSA, 15–16, 107–8, 229–31, 234; NYC-DSA goals relation to, 144–49, 159, 160, 161, 163–65, 177, 185, 193. *See also* moral failures
Sunkara, Bashkar, 68
Sunrise Movement, 198
Supreme Court, US, 241
SYRIZA, 54, 55

Tammany Hall, 27, 99
Tarleton, John, 73–74
Tax Justice Network, 196
"Tax the Rich" campaign (2019 and 2023), 65, 75–76, 161

Tea Party, 7–8
The Third Way (Giddens), 86
This Changes Everything (Klein), 32, 89, 204
Thunberg, Greta, 1, 33, 40, 44
Tilly, Charles, 58–59, 60
Trump, Donald, 4–5, 11, 43, 65, 231–32, 242; Brooklyn after 2016 victory of, 77, 109–10; NYC-DSA response to 2016 victory of, 77–79, 126, 133, 136, 138, 176, 198; racist ideology of, 77; socialism, use of term, 87. *See also* anti-Trump Resistance movement
Twitter/X, 84, 127, 134, 136, 139, 140
Tykulsker, Nadia, 116, 121, 125, 220

Uetricht, Micah, 21, 96–97
United Nations (UN), *IPCC 2018* report of, 200–201
United States (US), 241; democracy in, crisis of, 8, 13, 228, 232–33; Green Party in, 186, 196–97; welfare and democratic socialism in, 16, 94. *See also specific locations and organizations*
universalism, in democratic socialism, 6, 55, 83, 225, 232
UN report. *See* United Nations, *IPCC 2018* report of
US. *See* United States

Van Auken, Tascha, 156; on Democratic clubs, 183; on NYC-DSA and organizing approach, 178–82; on Obama campaign in 2008, 66–67, 74–75, 159–60; on Occupy Wall Street, 69–70; on Sanders compared with Obama campaigns, 74–75; on "seriousness" in organizing, 163, 165
vicious cycle, of declining participation, 31, 41–46
victories and losses. *See* successes and failures
visual identity: Alexis campaign in 2022 and, 141; cultural resources in

NYC-DSA, 122–23; Green New York design system, 224–25; millennial socialist, 83; movement culture, 225; NYC-DSA, 84, 115, 118, 122–23, 141, 146; organizational cohesion and, 158; stimulating participation, 122–23, 141
volunteers, 59, 75, 94, 95, 116; casual, 130–34; organization based on, 124; organizers, 137–40; pundits, 134–36; resources and burnout, 144, 158. *See also* leadership, in NYC-DSA; leadership theory; political organization/organizing
volunteers, in Alexis campaign, 122–24; casual canvassers and, 130–34; organizer-canvassers and, 137–40; pundit-canvassers and, 134–36

vulnerable populations, 193, 200–201, 202

Weber, Max, 49
welfare: decline in neoliberalism, 4, 5, 54, 86; European welfare states, 16, 55; New York City history of, 9; services and Democratic clubs, 99, 110; US democratic socialism and, 16, 94
Witthaus, Whitney, 93–95
Wood, Robert, 116–17, 119, 129
Wood, Ronin, 84, 141, 158
Working Families Party, 76, 187
World Economic Forum, 40, 52

X. *See* Twitter/X

Young Sheldon (television show), 87

ABOUT THE AUTHOR

FABIAN HOLT is Associate Professor in the Department of Communication and Arts at Roskilde University. He is the author of *Everyone Loves Live Music: A Theory of Performance Institutions*.

www.ingramcontent.com/pod-product-compliance
Lightning Source LLC
Chambersburg PA
CBHW020405040426
42333CB00055B/469